IMAGES
of Aviation

EARLY AVIATION IN LONG BEACH

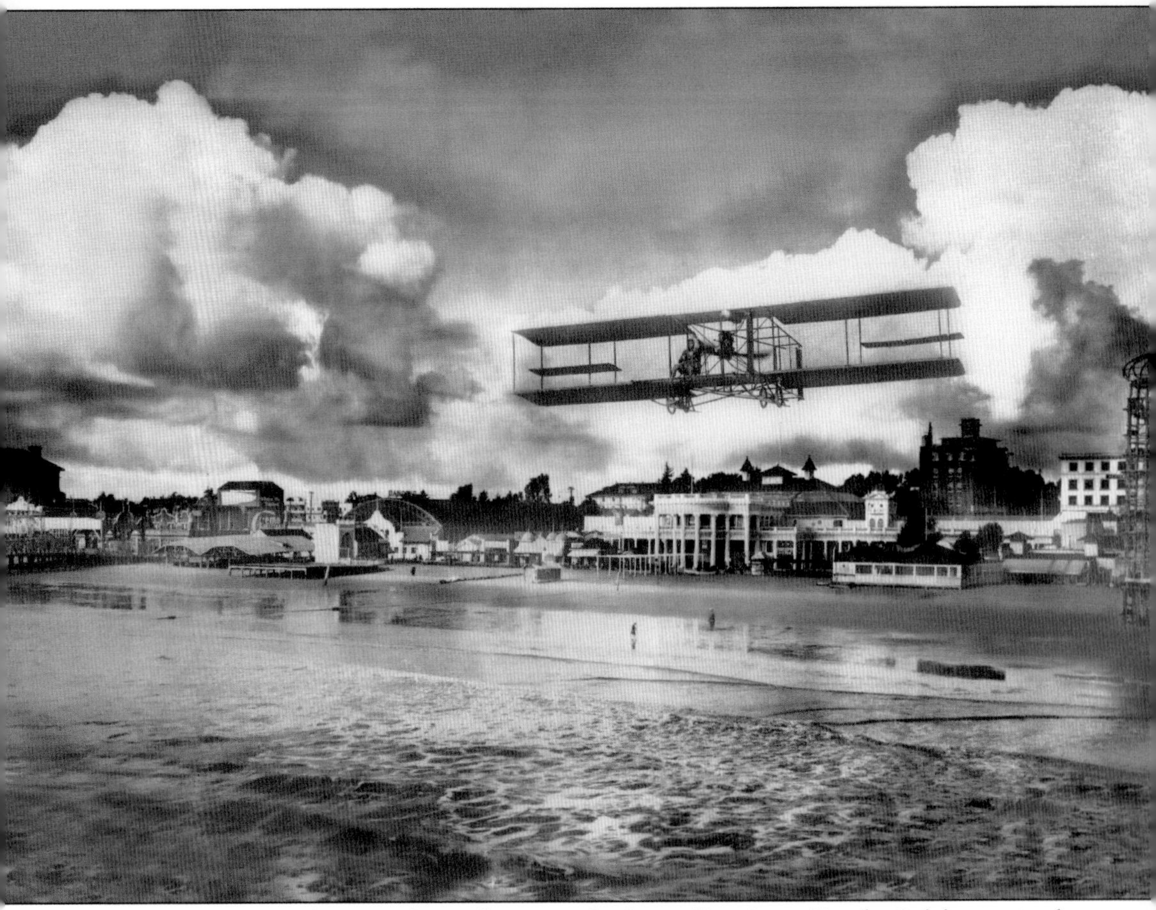

On the Cover: Earl S. Daugherty flies over the "strand" in Long Beach, California, in the plane he built in the early 1900s. Daugherty established the city's first non-beach airfield in 1919. (Courtesy Long Beach Municipal Airport Archives.)

IMAGES
of Aviation

EARLY AVIATION IN LONG BEACH

Gerrie Schipske

ARCADIA
PUBLISHING

ISBN 978-0-7385-7083-9

Published by Arcadia Publishing
Charleston, South Carolina

Printed in the United States of America

Library of Congress Control Number: 2009924707

For all general information contact Arcadia Publishing at:
Telephone 843-853-2070
Fax 843-853-0044
E-mail sales@arcadiapublishing.com
For customer service and orders:
Toll-Free 1-888-313-2665

Visit us on the Internet at www.arcadiapublishing.com

This book is dedicated to the women and men who made the Long Beach Municipal Airport–Daugherty Field the finest in the nation.

Contents

ACKNOWLEDGMENTS

A special acknowledgment must be given to Ken Larkey for his long-term efforts to preserve and promote the aviation of history of Long Beach. Special thanks go to the Long Beach Municipal Airport Public Affairs Department for their patience and assistance.

Unless otherwise noted, all images appear courtesy of the Long Beach Municipal Airport Archives (LBMAA). The other images in this volume appear courtesy of Library of Congress (LOC), National Archives (NA), Boeing Company (BC), U.S. Naval Historical Collection (USNHC), Long Beach Transit (LBT), author's collection (AC), California State University Dominguez Hills (CSUDH), R. T. Gerow Collection (RTG), Long Beach Municipal Library Archives (LBMLA), Pat Loop, City of Glendale Library Special Collections (GLSPC), Ken Larkey (KL), Hawaii Aviation Preservation Society (HAPS), and www.flightglobal.com.

Introduction

At the dawn of the 20th century, several forces drove the growth of aviation in Southern California: climate, daring pilots, the railroads, automobiles, the motion picture industry, oil, ports, aircraft manufacturers, and the military.

Long Beach, California, had them all. Long Beach was both an ideal location for early aviation and a land speculator's dream. The seaside town quickly grew as developers marketed its assets: "the longest beach drive, the finest surf-bathing and the best fishing on the Pacific Coast." Advertisements promised one could "make millions" by investing in the land.

Soon the town had boardinghouses, hotels, parks, schools, a newspaper, and a pier and wharf that jutted out thousands of feet over the Pacific Ocean. Its Chautauqua summer camp and conservative Methodist Assembly brought thousands to visit each summer.

The San Pedro, Los Angeles, and Salt Lake Railroad (SPLA&SL), which eventually became Union Pacific Railroad, ran from Los Angeles to San Pedro and routed through Long Beach to bring thousands of visitors. Railroad moguls and developers built an electric trolley system (Pacific Electric Railway) in 1902 to bring potential investors from Los Angeles and Pasadena to an area adjacent to the ocean called "The Walk of the Thousand Lights." The area included a large amusement site ("The Pike") and bathhouse. The Pike featured games, movie and live theaters, skating rinks, sideshows, Bisbey's Spiral Airship, and other aviation-themed rides. Sand lots offered in 1903 for $100 sold in 1904 for $500.

As early as 1905, the strand of sand in front of the Pike and adjacent to the pier provided a soft landing spot for air balloons used to attract visitors. The beach served as Long Beach's first takeoff and landing area for "heavier than air" machines until a young man from Des Moines, Iowa, Earl Stanley Daugherty, leased land in 1919 and then convinced the City Council to establish Southern California's first municipal airport in 1923. Today it is known as the Long Beach Municipal Airport–Daugherty Field.

Seven years after the Wright brothers tested their "Wright Flyer" off the sand dunes of Kitty Hawk, North Carolina, the first U.S. international air meet was held January 10–20, 1910, at Dominguez Field just steps away from Long Beach. Daugherty and three other Long Beach residents—Frank Champion, Frank Stites, and Harry Holmes—caught the "flying bug" and by 1911 became the first Long Beach residents to qualify for an international license from the Aero Club of America, an affiliate of the Automobile Club of America, which sanctioned competitive flying events. (Champion received a license in 1910 while taking flying lessons at the Bleriot Aviation School in Hendon, England, making him the first Long Beach resident to receive a license to fly.)

The Long Beach Chamber of Commerce sponsored the Long Beach Trophy for altitude at the second Los Angeles international air meet at Dominguez Field, held December 24, 1910–January 3, 1911. The trophy was won by Arch Hoxsey, who reached a record of 11,474 feet.

In 1911, Calbraith Perry ("Cal") Rodgers completed the first U.S. transcontinental flight from Sheepshead Bay, New York, to the surf off Long Beach near the Pine Avenue pier. Rodgers was to end his flight in Pasadena, but the chamber of commerce enticed him with $1,000 to finish in Long Beach. As he approached the city, his two friends Frank Champion and Earl Daugherty flew to meet him as he passed Dominguez Field and gave him an escort onto the long beach. Rodgers's flight is so significant in aviation history that a replica of his plane, the "Vin Fiz," is on display in the Smithsonian Air and Space Museum in Washington, D.C.

Both Daugherty and Champion built airplanes in the basements of the Virginia Hotel and the Earl, an apartment building located at 425–431 East Seaside Way on the sand owned by Daugherty's parents. Daugherty and Champion rolled out the planes to the sand and joined other flyers in performing aerial stunts over the beach. Their flying and plane-building skills were legendary and took them across the country to build and test planes and engage in numerous air contests.

Daugherty enlisted in the military when the United States entered World War I and served as a flying instructor in the Aviation Section of the Signal Corps, U.S. Army, both at North Island in San Diego and March Field in Riverside.

The U.S. Army and Navy took a great interest in Long Beach as early as 1907 after city officials traveled to Washington, D.C., and lobbied for a naval base, telling the military of Long Beach's "harbor advantages," offering "to donate the site if necessary."

In 1908, Pres. Theodore Roosevelt sent the 2nd Division battleship fleet painted white—"The Great White Fleet"—around the world with a stop and a barbecue attended by 50,000 in Long Beach. Daugherty's later connections with the military and his frequent invitations to military flyers (many of whom he trained) to participate in the Long Beach air races he organized helped solidify the military's relationships with Long Beach.

The years between 1918 and 1938 are called the "Golden Age of Aviation." Airplanes developed from flimsy flying machines to weapons of war and modes of transportation. When World War I ended, a surplus of military planes (Curtiss JN-4s, also called "Jennys") and trained flyers were available for civilian use. The Roaring Twenties and the growing silent movie industry promoted (and documented) the excitement of stunt-performing "barnstormers," who flew at every outdoor public event. Commercial investors offered lucrative cash prizes for speed contests, and several flyers (Lindbergh and Earhart) made aviation "firsts." Cities began considering the airport as a municipal utility.

Daugherty's skills and the fact that he hailed from Long Beach brought much national press attention to the city. Daugherty returned to Long Beach in 1919 and used a sand and sod field near American Avenue (now Long Beach Boulevard) and Bixby Road. He named the field Château Thierry Flying Field, which is the site in France of the first battle in which American troops were supported by airplanes and where many of his flying students died. The field was the city's first non-beach airfield.

The promise of oil and its discovery in 1921 near Signal Hill brought Long Beach incredible wealth, which it used to improve its harbor, thereby attracting more shipbuilding and the navy. Henry Ford built his 35th assembly plant on Terminal Island in 1930. The oil boon increased population and demand for housing and schools. Daugherty needed more space for his aviation activities and with the encouragement of the city council moved to 23 acres at Long Beach Boulevard and Willow Street (owned by his parents), where he hoped the new site would become a large air terminus. He ran a flying school and continued building planes.

Daugherty also sold planes, gave passengers rides, did police work from his Nieuport Speed Scout plane and organized "air tournaments," "air rodeos," and "air circuses" featuring wing walking, Jenny races, upside-down flying, and fireworks. The events attracted thousands, including a young Amelia Earhart, who in 1920 came with her father to watch Daugherty's stunt flying. After the event, she asked to be taken up in a plane. Long Beach Poly High School graduate Frank Hawks obliged Earhart the next day, and the history of aviation was forever changed.

Daugherty met John G. Montijo in the Army Air Corps. Later Montijo tested airplanes at and around Long Beach and taught aerobatics. His business partner, W. B. Bert Kinner (owner and operator of Kinner's Field and Kinner's Airplane and Motor Corporation), asked Montijo to take over teaching Amelia Earhart how to fly. Earhart had taken a few lessons from another Kinner

Field flyer, Neta Snook, but it was Montijo who taught Earhart how to solo and certified her in advanced aeronautics and aerobatics. Shortly after, Earhart performed with Daugherty, Frank Clarke, Frank Hawks, and Wesley May in the Long Beach Air Rodeo.

November 12, 1921, Frank Hawks and Early Daugherty performed the first ever midair refueling. Long Beach stunt-walker Wesley May climbed from Hawks's plane onto Daugherty's with a 5-gallon can of gasoline strapped to his back and then "replenished an imaginary failing supply" of fuel. He then stood on his head on the plane's upper wing. Fellow stunt-walker Freddie Lund performed a similar stunt between planes piloted by Frank Tomick and Frank Clarke. Midflight refueling became a critical maneuver with military aircraft in World War II. The 1921 Hawks-May-Daugherty event is featured as a painting in the Leach International Heritage of the Air Collection of key events in aviation history.

In 1923, the city council dedicated 80 acres of water department land at Cherry Avenue and Spring Street (north of Signal Hill) for the first municipal airport in Southern California, the Long Beach Municipal Airport. The council passed an "aviation ordinance" in 1924 and approved the navy's use of the municipal airport for its fleet planes. Daugherty, Montijo, and A. R. Ebrite were named to the city's first Aviation Commission in 1924.

Daugherty, who received a commission in the navy as a lieutenant, junior grade, invited flyers to his May 1924 "air rodeo." After 11 navy planes participated, the navy exclaimed, "The Long Beach municipal aviation field at Spring Street and Cherry Avenue is one of the best in southern California. . . . the Long Beach field is remarkable, in that it is a four-way field, or a field where it is possible for aviators to land toward any quarter of the compass. It is possible to land on many fields in one way only."

John Montijo and his business partner and local Dodge dealer C. B. "Boone" Bellows signed the first lease on the airport in 1925. They built, then sold, a hangar to Earl Daugherty for his flying school, passenger service, and airplane sales office. That year, Daugherty set a record of carrying 7,000 passengers from Long Beach Municipal Airport without a single mishap.

Fourteen additional hangers were added to the airport to accommodate aviation schools, airplane manufacturing and repair, and surveying services.

In 1926, T. "Tubal" Claude Ryan, a barnstormer who trained at March Field, opened mail carrier service at Long Beach. Ryan's company used the Douglas Cloudster that launched the manufacturing career of Donald Douglas. Ryan Airlines produced the Ryan NYP, commissioned by a group of investors and renamed *Spirit of St. Louis.* (Douglas Corrigan worked on the assembly of the plane and later tried to re-create Lindbergh's success but feigned he was flying "the wrong way" when he landed in Ireland instead of Long Beach.)

Charles A. Lindbergh flew the *Spirit of St. Louis* on his historic trans-Atlantic flight on May 20, 1927. He also came to Long Beach that year and, while there, asked aviator local Lloyd O'Donnell to make some minor repairs on his plane. O'Donnell couldn't resist and took several Long Beach residents up in the famous plane, charging $60 per person.

Lindbergh visited several more times, including during the 1928 Pacific Southwest Exposition hosted by Long Beach that attracted hundreds of thousands of visitors. Later he was able to make an emergency landing after midnight with his passenger Col. Henry Breckenridge—only because a resident heard the plane overhead and telephoned the field and a light switch was turned on, guiding Lindbergh to the field. A grateful Lindbergh returned to Long Beach and publicly lauded the city for its state-of-the-art features, which included the light and paved runways. He stayed frequently at the Breakers Hotel on Ocean Boulevard.

After the crash of Earl Daugherty's plane in late 1928—killing him, Warren Monfort (city editor of the *Press-Telegram*), and Elmer Starr (one of the owners of the Pacific Engraving Company)—the city renamed the airport in his honor: Long Beach Municipal Airport–Daugherty Field.

Many women were active in aviation in Long Beach. Earhart performed flying stunts in Daugherty's air rodeos. Gladys O'Donnell, Long Beach's first licensed woman pilot, competed against Earhart in the 1929 Women's National Air Derby, placing second. Earhart finished third. Long Beach sponsored the 1930 Women's Air Derby, which ended in Chicago, with O'Donnell capturing the win. O'Donnell was a founding member (with Earhart) of the Ninety-nines, the first organization of female pilots. She later served on the city's Aviation Commission and in the leadership of the Republican party.

The city built hangars and facilities for the navy and the Army Air Corps at the airport in the late 1920s but aggressively pursued commercial aviation and made it clear it preferred non-military use of the airport. By 1929, Western Air Lines and Maddux Airlines carried commercial passengers, mail, and cargo from the Long Beach Municipal Airport.

Oil money allowed the city to make drastic improvements and expansion of its oceanfront, harbor, and airport in preparation of becoming a major commercial site in Southern California. Although never officially approved by the city council, the design and layout of the airport is credited to the vision of the city engineer of Long Beach.

In the late 1930s, the council approved plans to purchase 255 acres adjoining the municipal airport and the construction of a three-story administration building and tower at the east side of the airfield. The airfield was improved with 1,000 WPA workers, and local press remarked, "Long Beach airport, which will be one of the biggest and best in the country, faces a great future as a magnet for industrial payrolls."

The military eyed Long Beach as a site for an aircraft manufacturing center in the late 1930s because of its developed airport and access to the port and railroad. Douglas Aircraft opened adjacent to the airport in 1941. World War II brought the Army Air Corps 6th Ferrying Group Air Transport Command to Long Beach, and the airport became the center of ferrying all military aircraft produced in Southern California.

Female civilian pilots flew thousands of the military planes assembled locally to other military bases. Known as Women's Auxiliary Ferry Squadron (WAFS) and the Women Airforce Service Pilots (WASP), some were attached to the 6th Ferrying Group and commanded by Barbara Erickson London. During the war, numerous movie stars and government dignitaries, such as President and Mrs. Roosevelt, paid visits to the Long Beach airfield.

Throughout the war, Long Beach continued planning for commercial aviation, visualizing a 2,340-acre airport after the war. The terminal and control tower building was completed and scheduled for grand opening on December 7, 1941, the day Pearl Harbor was attacked. Instead, it was repainted in camouflage and used as a billet for soldiers and military guns. The terminal formally opened on April 26, 1942 (and its pastel paint was restored in 1945). Designed by Horace Austin and Kenneth Wing in the streamline moderne style of architecture, the building (still) looks more like a ship than an airplane terminal. Inside were the mosaics and murals of WPA artist Grace Clements, which won national acclaim. Long Beach declared the terminal (mostly unchanged since 1941) a cultural landmark in 1990. It remains the only terminal at the Long Beach airport.

The continued presence of the military until the late 1950s, the large number of private planes flying from the field, and the rapid development of residential housing around the field eventually prevented Long Beach from competing with Los Angeles to become the terminus for large airlines.

Today most of the other California airports begun in the golden age of aviation have closed. Long Beach Municipal Airport–Daugherty Field continues to be an active center of Southland aviation. Occupying 1,166 acres, it is among the top five busiest general aviation airports in the world with 365,000 annual general aviation operations. It is also a center for air cargo carriers and hosts several commercial airlines.

One

Early Long Beach

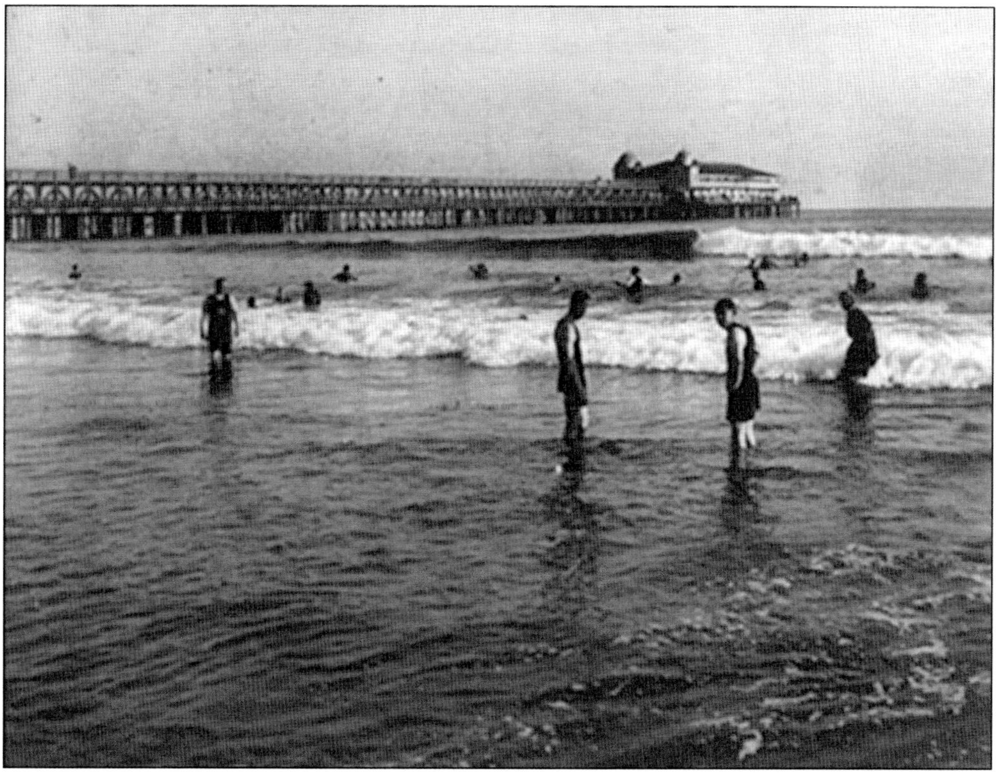

Land Speculator's Dream. Long Beach began as a project of the California Immigrant Union (CIU), organized by William Erwin Willmore as the American Colony. Willmore purchased part of Rancho Los Cerritos from J. Bixby and Company. In 1884, Willmore abandoned the land, which was subsequently purchased by the Pomeroy and Mills real estate investors and renamed Long Beach at the suggestion of Belle Lowe, the wife of the city's first postmaster. (LOC.)

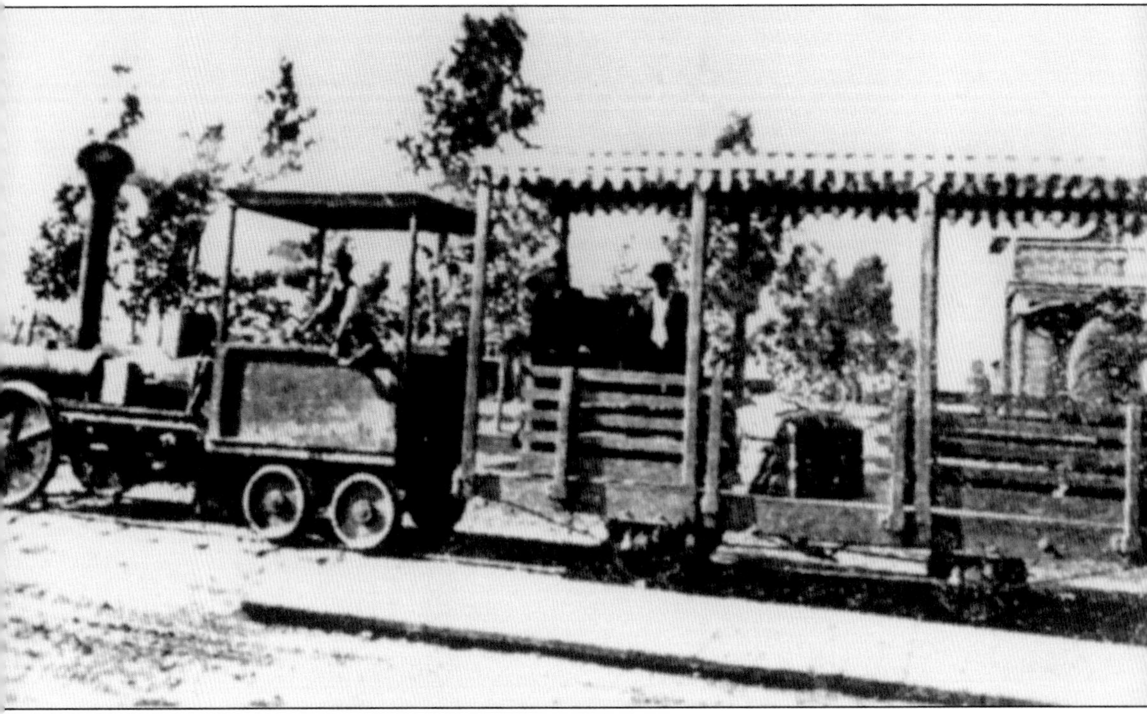

Get Out and Push Railroad. Judge Whitney established a horsecar line to bring tourists from Los Angeles and Wilmington to Alamitos Avenue and Ocean Boulevard. The pine and redwood rails sagged under the weight of the cars, requiring passengers to get out and push the cars. Eventually replaced with a steam engine, it too had had problems that required the engineer to yell to the passengers "Get out and push!" The problem was fixed in 1887 when the line was purchased by the Southern Pacific Railroad. Electricity provided by the Edison steam plant soon powered a trolley. (LBT.)

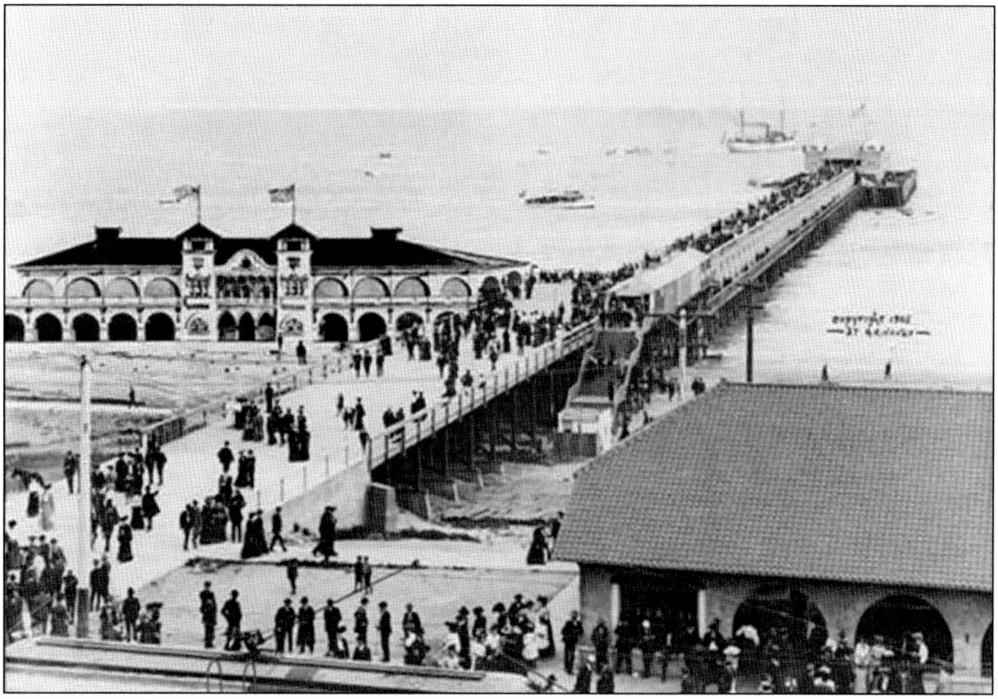

The 17,000-Foot Pier. Extending 17,000 feet over the ocean, the pier attracted thousands including anglers, many of whom fished at night because of a powerful submarine electric globe, probably the only one of its kind on the Pacific Coast, suspended into the water at the side of the wharf and beyond the breakers. The pier collapsed in 1913, killing 33 after 10,000 crowded into events celebrating British Empire Day. (LOC.)

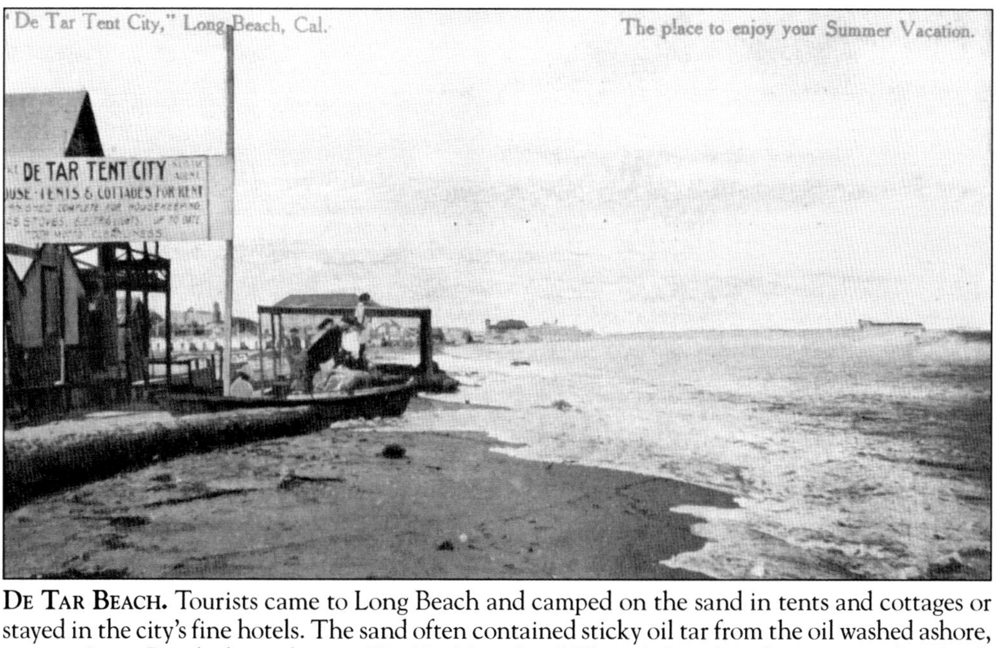

De Tar Beach. Tourists came to Long Beach and camped on the sand in tents and cottages or stayed in the city's fine hotels. The sand often contained sticky oil tar from the oil washed ashore, earning Long Beach the nickname "De Tar Tent City." The oil, found in the ocean and on land, would later prove to be a major source of Long Beach's wealth. (AC.)

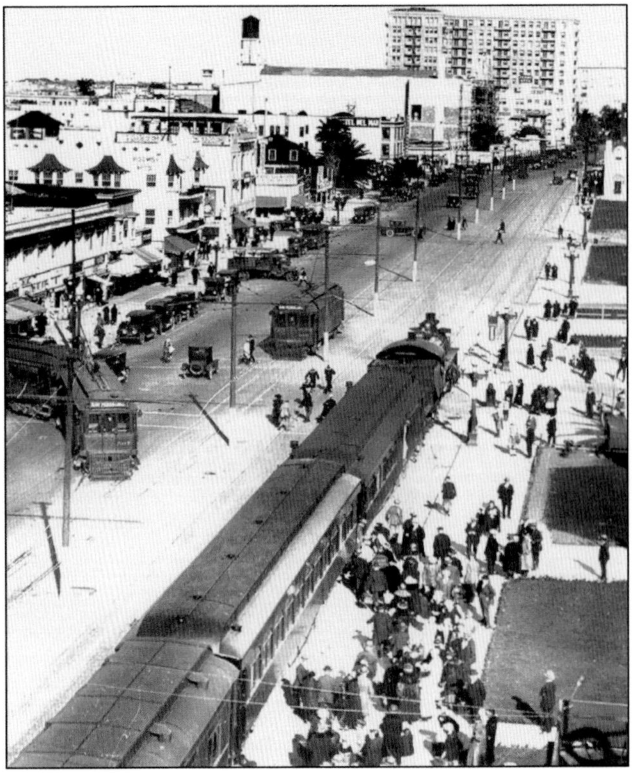

Train and Trolley. Long Beach real estate developers and railroad moguls routed trains and trolleys to the city. The Union Pacific Railroad from Los Angeles to San Pedro came through Long Beach. Col. Charles R. Drake, president of the Long Beach Bathhouse and Amusement Company, brokered the franchise to build the Pacific Electric trolley to Long Beach on behalf of Henry E. Huntington. Drake also promoted the development of the bathhouse and amusement area called the Walk of the Thousand Lights (also known as the Pike) as well as the Bixby Hotel, which collapsed in 1905 and was rebuilt as the Virginia Hotel. Drake headed the Seaside Water Company, which purchased the Bouton Water Company and the Long Beach Development Company. (Both, LBT.)

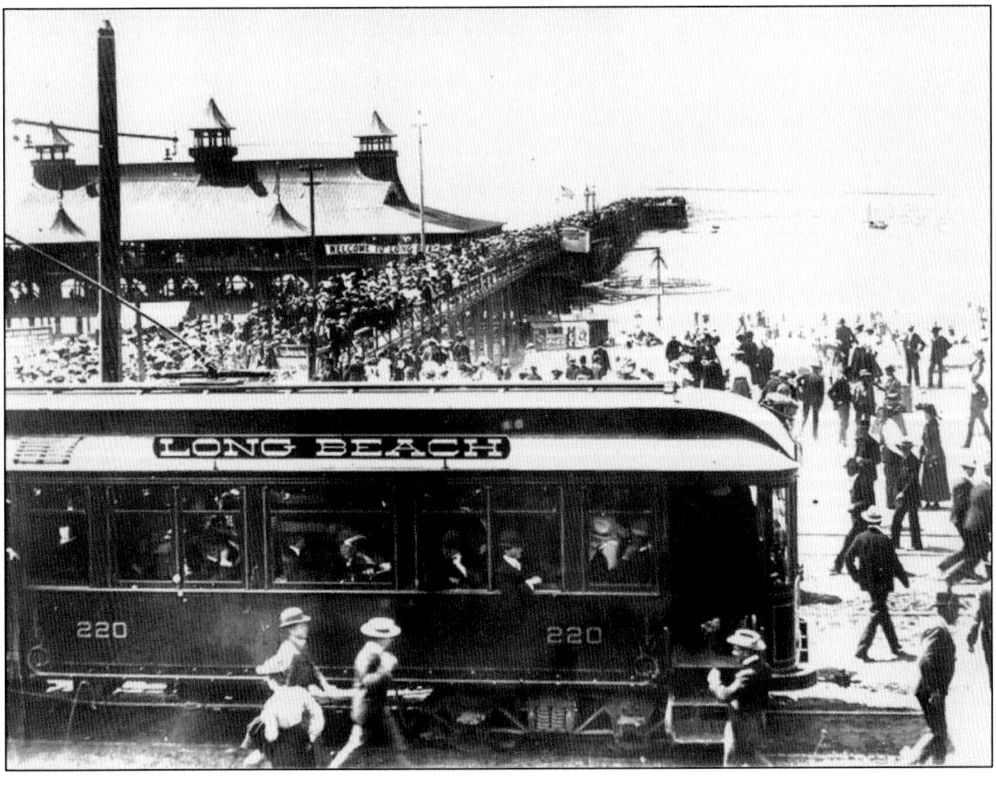

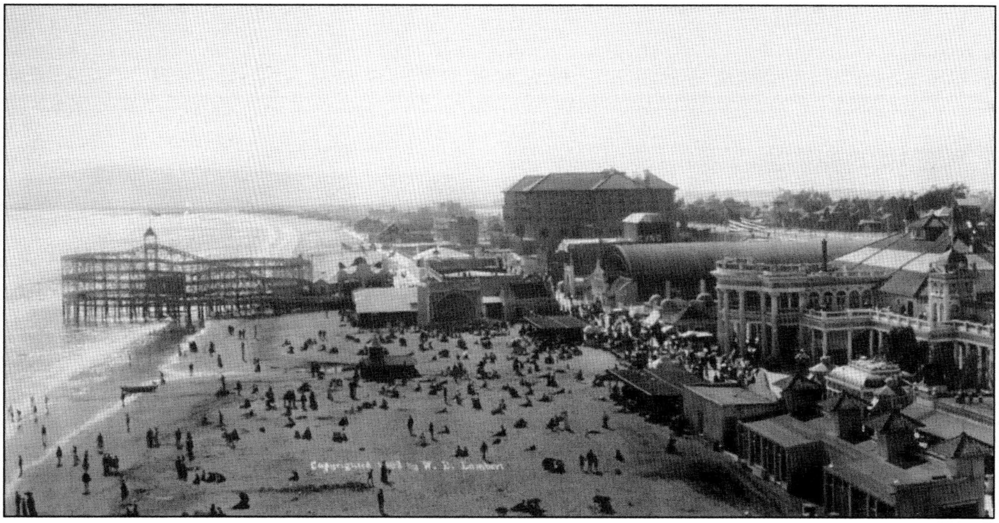

The Pike. The bathhouse (the building with columns on the right) and the amusement area known as the Pike and the Pacific Electric trolley opened in 1902, bringing thousands of visitors and prospective buyers to Long Beach. Sand lots offered for sale in 1903 for $100 sold for $500 by 1904. (LOC.)

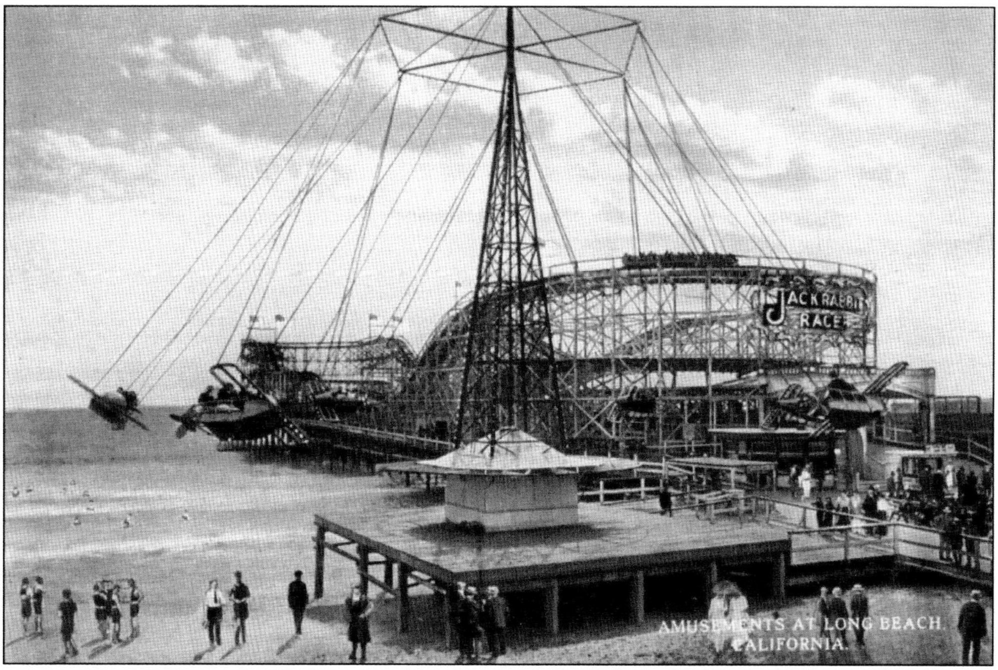

Aviation Rides. The Pike featured carnival games, movie and live theaters, skating rinks, sideshows, rollercoaster, Bisbey's Airship, and other aviation-themed rides that allowed riders to soar over the sand. Real airplanes also looped around the roller coaster and soared over the sands in Long Beach until its first in-land airfield was established by Earl S. Daugherty in 1919. (AC.)

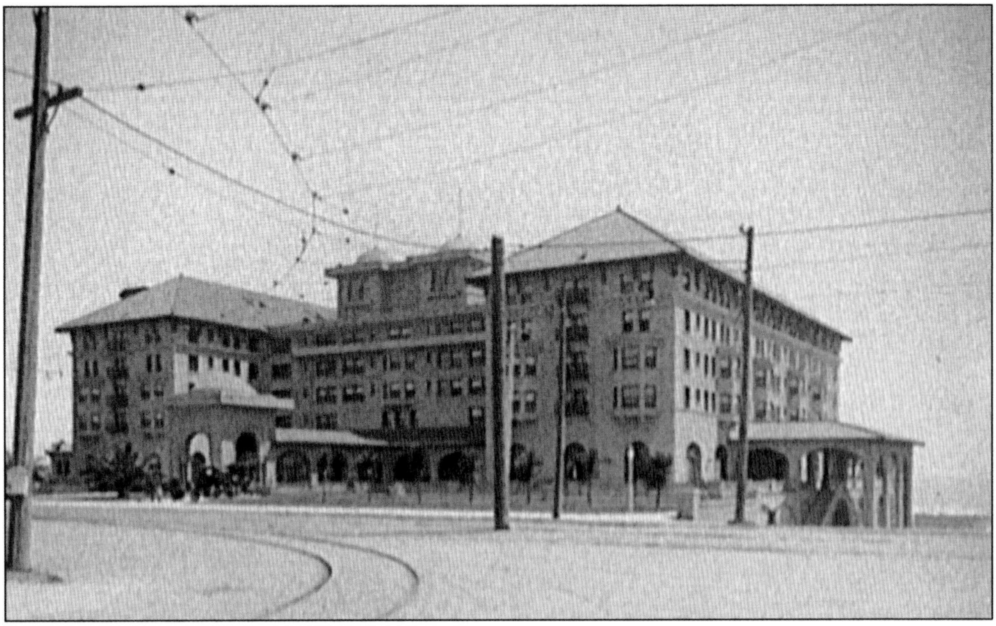

VIRGINIA HOTEL. The city advertised its elegant Virginia Hotel, whose basement became the perfect site to construct an airplane and roll it out to the sand. The Virginia Country Club was developed to accommodate the guests of the Virginia Hotel. The club was initially located in the area known as Recreation Park and in 1921 relocated to its current Los Cerritos site. (LOC.)

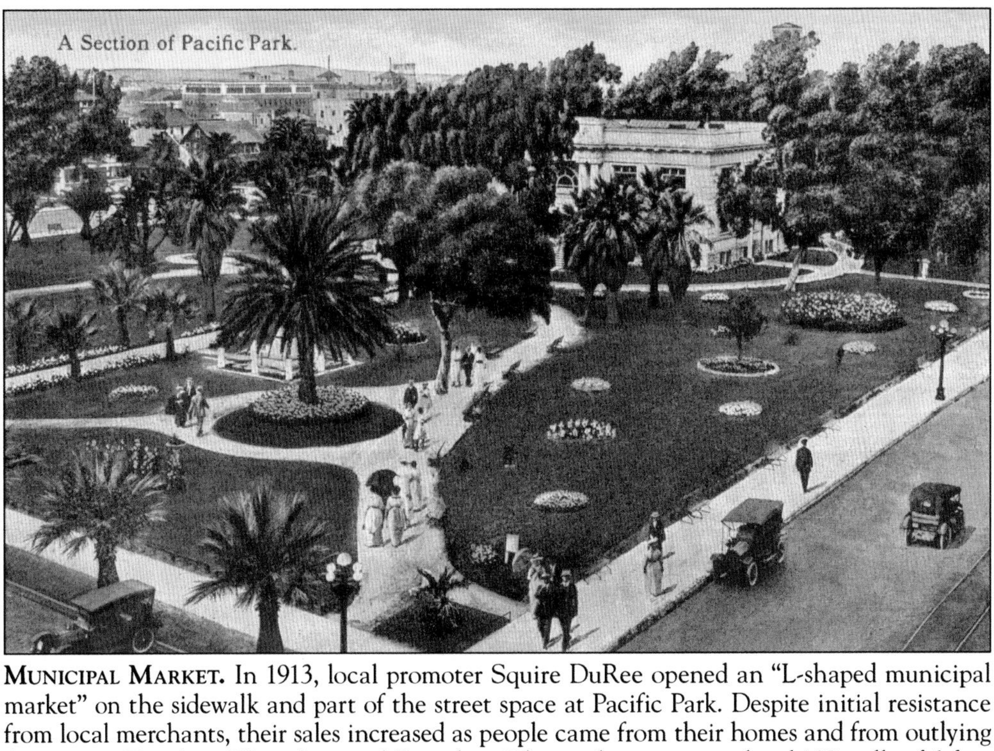

MUNICIPAL MARKET. In 1913, local promoter Squire DuRee opened an "L-shaped municipal market" on the sidewalk and part of the street space at Pacific Park. Despite initial resistance from local merchants, their sales increased as people came from their homes and from outlying districts on Tuesdays, Thursdays, and Saturdays. The market accommodated 109 stalls of 6-foot frontage, 18 feet deep, and brought as many as 7,000 shoppers. (AC.)

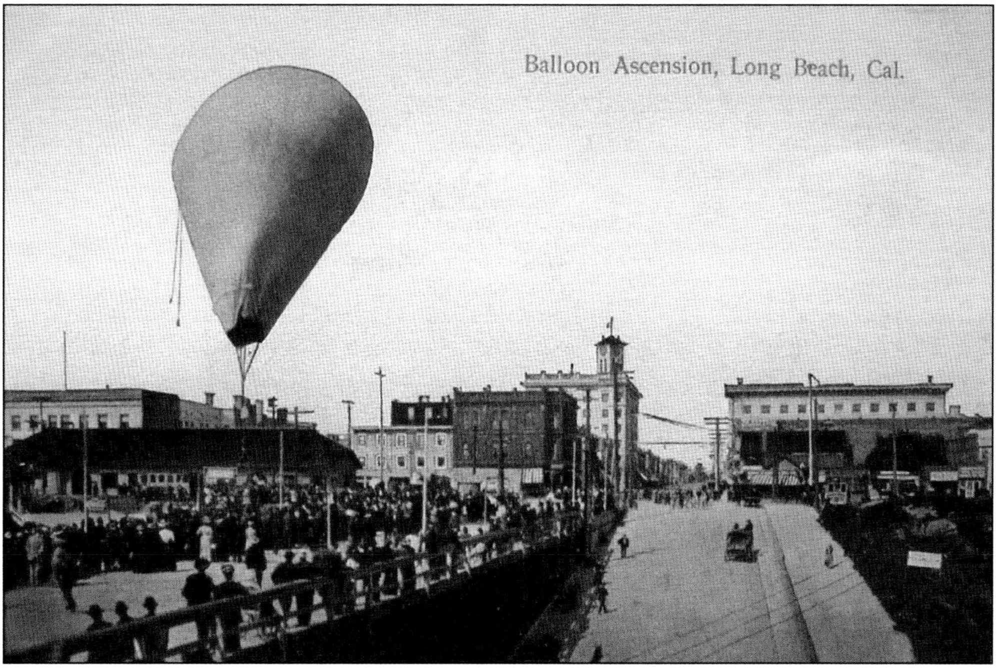

Balloons and Airplanes. Promoters of the pier utilized air balloons to draw large crowds as early as 1905. A few years later, airplanes soared over both the pier and the municipal market, dropping stunt performers onto the beach. Stunt flyer Frank Stites swooped over the sand near the Pike in 1908. Stites, who earned his pilot's certificate in 1911, became a stunt pilot for Universal films and died in 1915 when an explosive used for a special effect blew him out of his cockpit. (Both, AC.)

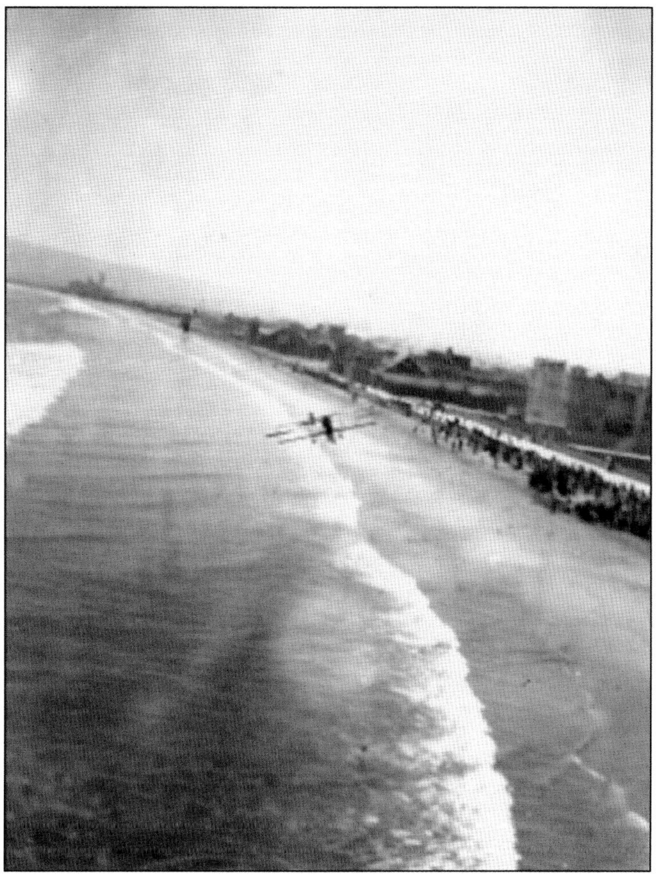

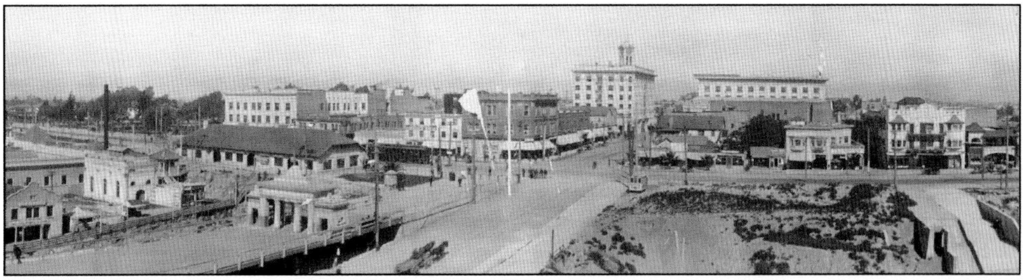

BEST INVESTMENT IN THE WEST. By 1909, the *Long Beach Press* boasted of city assets: "Long Beach . . . is a cosmopolitan city of about 20,000 people. It has nine miles of crescent beach, a $4,000,000 hotel open all the year, a $250,000 shipbuilding plant, $100,000 pleasure pier, $35,000 auditorium overhanging the ocean (seats 6,000), 12 miles of paved streets, 72 miles of cement curbs and walks, sewage system costing $201,000, building permits 1906–08 $2,564,095, nine magnificent schools, 30 churches, four national banks with deposits of $3,000,000. Long Beach is the fastest growing city in the United States. Population 1900 was 2,173; 1909, 20,000. Watch it grow." National magazines promised speculators that "Long Beach Estate" was "The Best Investment in the United States." (Above, LOC; below, AC.)

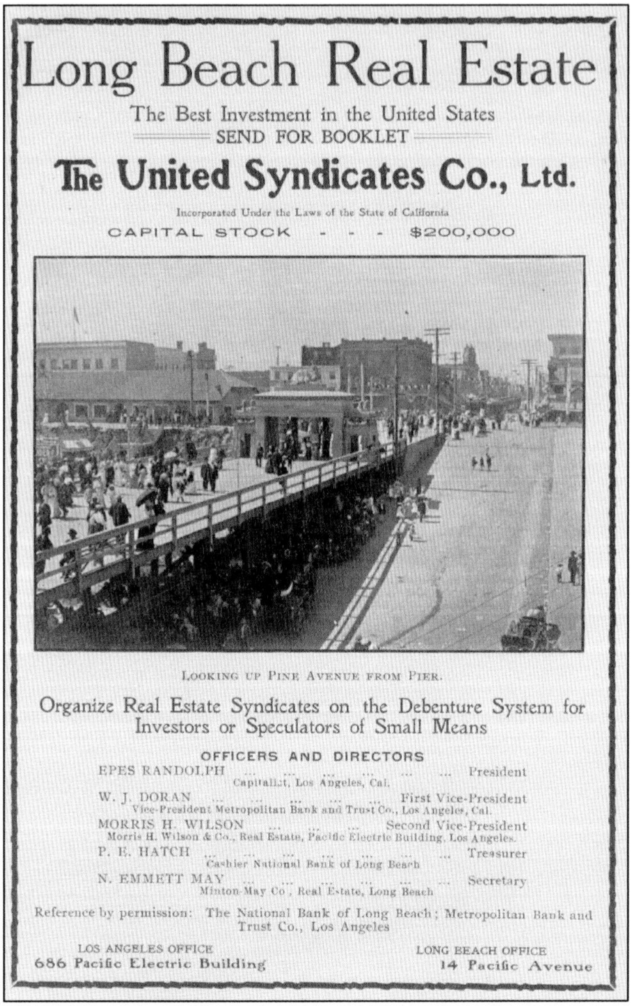

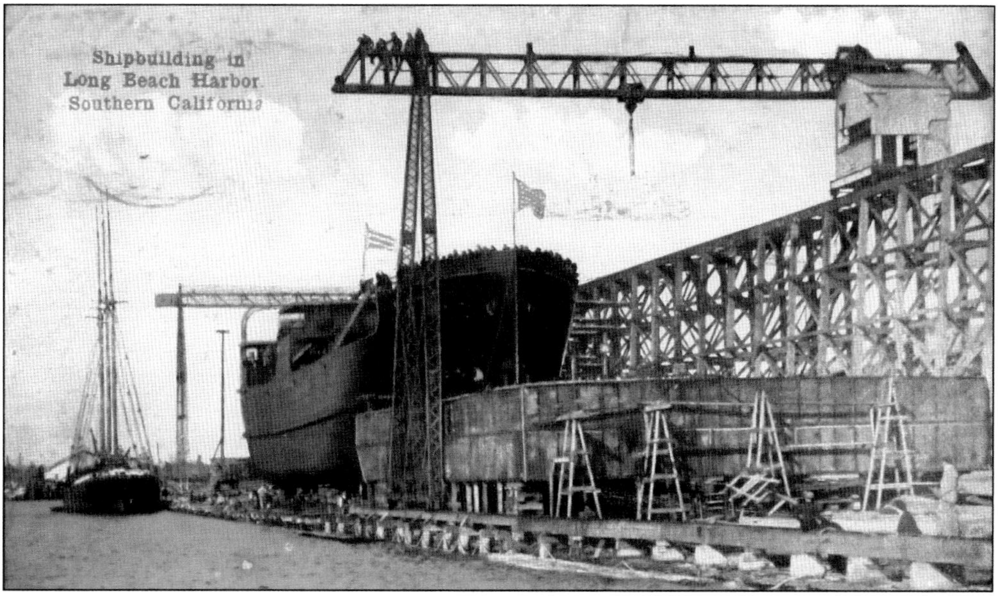

SHIPYARD AND SUBMARINES. Port development brought major manufacturers of ships, submarines, automobiles, and airplanes. While Collis Huntington of the Southern Pacific Railroad fought for a deepwater port off Santa Monica, Long Beach helped San Pedro to become a federal harbor. The Long Beach port was opened in 1907 by the Los Angeles Dock and Terminal Company, which owned 800 acres of mud flats to the west of the city. Craig Shipyards relocated from Toledo, Ohio. The Panama Canal opening in 1914 brought international shipping and passenger steamboats directly into Long Beach. By 1916, the city operated the Dock and Terminal Company, and by 1918, the U.S. Army Corps of Engineers dredged the Cerritos Channel, opening navigation between the San Pedro and Long Beach harbors. (Above, AC; below, USNHC.)

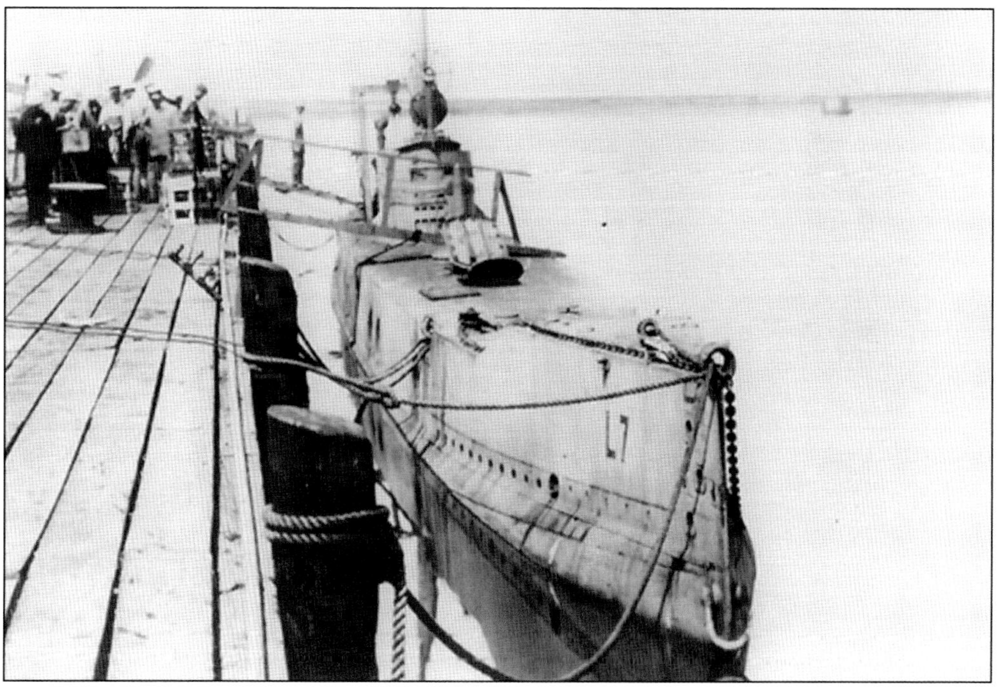

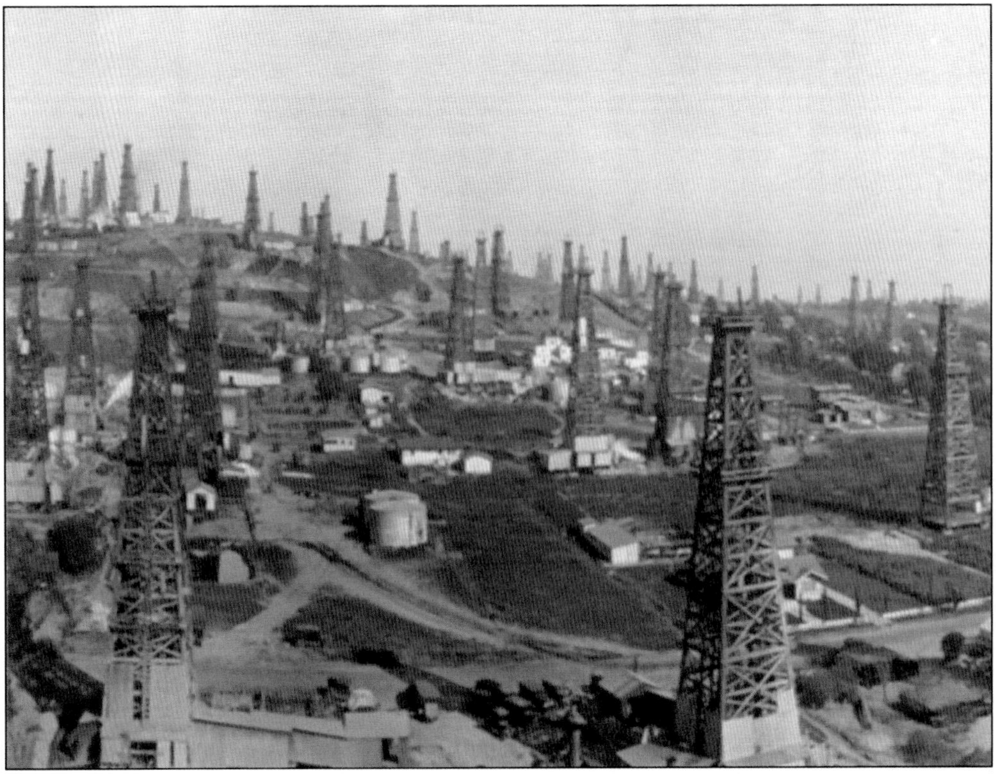

Oil Discovered. In 1909, Long Beach expanded to Belmont Heights; in 1910, it acquired 3.8 square miles north of the city; in 1911, it sealed off Los Angeles from Long Beach by acquiring Alamitos Bay and Naples and land in Signal Hill. On June 23, 1921, Shell Oil's Alamitos No. 1 well hit, bringing Long Beach a source of revenue that was used to invest in public works projects, such as establishing the first municipal airport in California. The city expanded to the Zaferia area, Los Cerritos, Bixby Heights, and an area called Château Thierry. The city's population grew to over 100,000 by 1924. The city was advertised as "Queen of the Beaches." Continued growth in population would be a source of labor for the Ford plant (opened in 1930) and Douglas Aircraft plant (opened in 1941). (Both, LOC.)

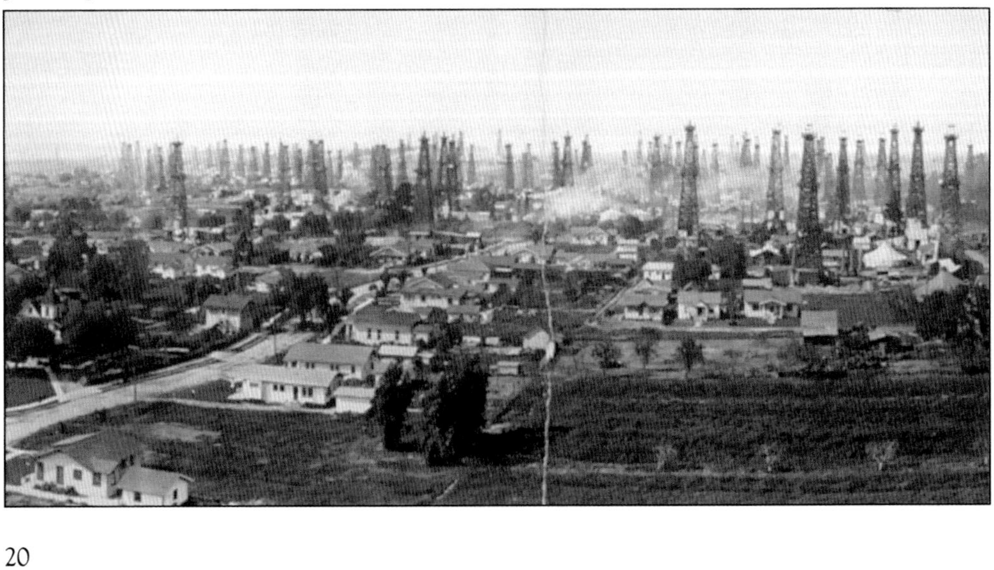

Two

AVIATION TAKES OFF IN AMERICA

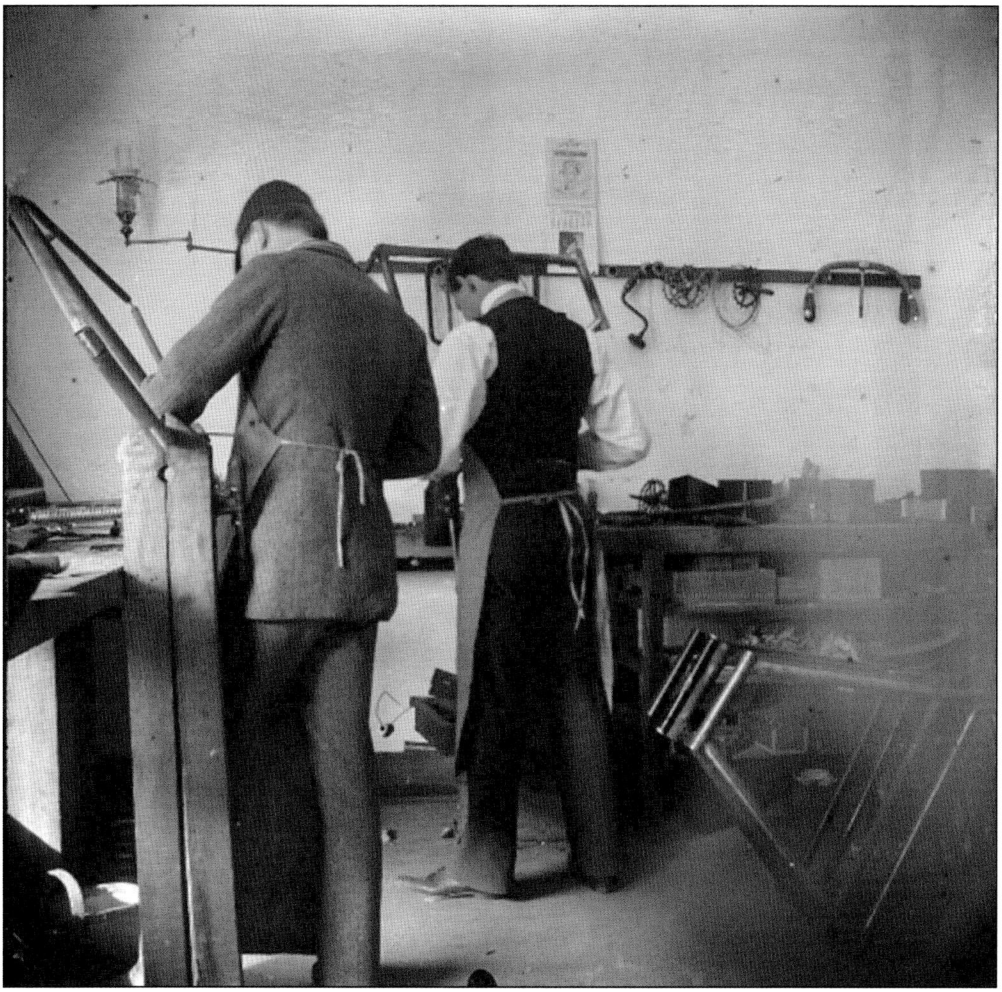

WRIGHT BROTHERS. In 1903, Orville (pictured left with his friend Edward Stines) and Wilbur Wright, brothers who owned a bicycle shop (pictured), tested their "Wright Flyer," an engine-propelled, heavier-than-air flying machine, off the sand dunes of Kitty Hawk, North Carolina, proving that man could indeed fly. (LOC.)

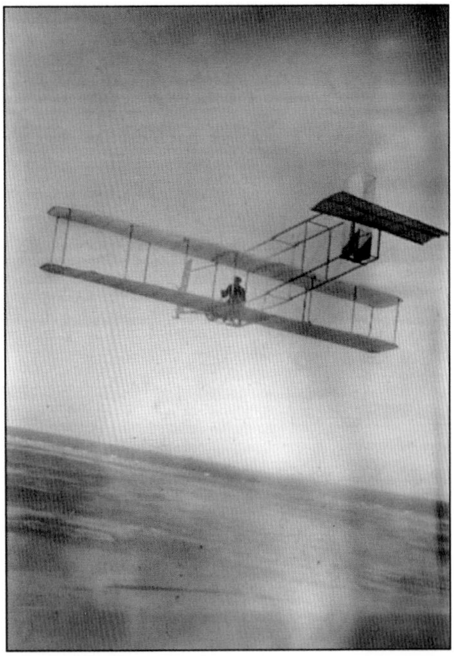

LIGHTER THAN AIR. The Wright brothers' flight was quickly copied. The Wrights sued anyone who attempted to infringe on their design, stalling aviation here while it exploded in Europe. The 1909 Grande Semaine d'Aviation flying contest organized by the Aero-Club de France at Rheims, France, drew the world's greatest European and American aviators, balloonists, and dirigible operators, such as Glenn Curtiss, Lincoln Beachey, and Roy Knabenshue. (LOC.)

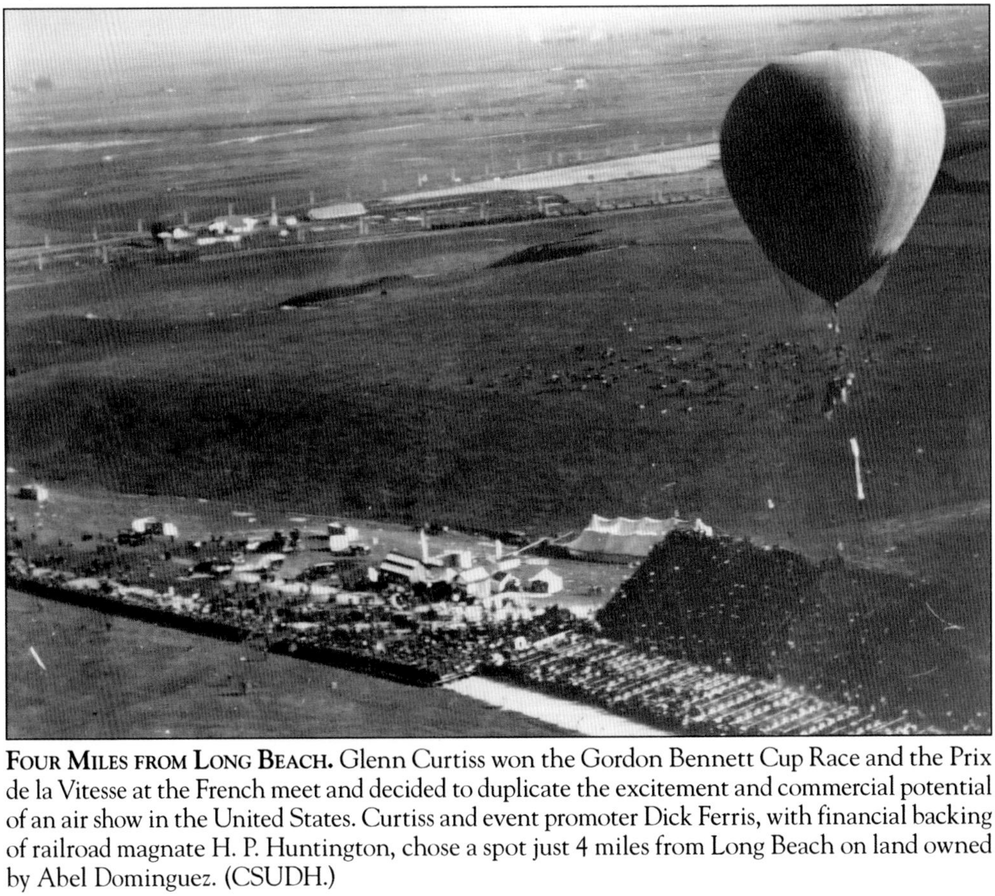

FOUR MILES FROM LONG BEACH. Glenn Curtiss won the Gordon Bennett Cup Race and the Prix de la Vitesse at the French meet and decided to duplicate the excitement and commercial potential of an air show in the United States. Curtiss and event promoter Dick Ferris, with financial backing of railroad magnate H. P. Huntington, chose a spot just 4 miles from Long Beach on land owned by Abel Dominguez. (CSUDH.)

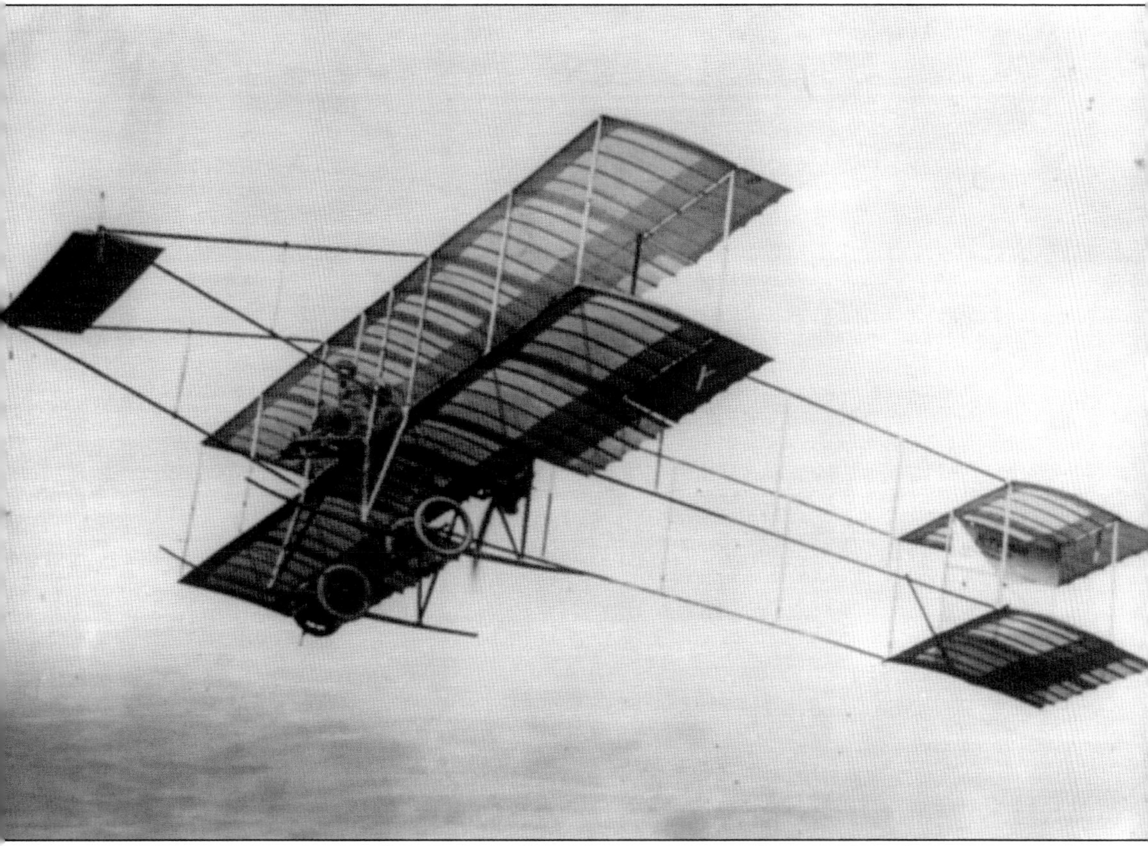

Dominguez Field. Louis Paulhan (pictured) was among many who soared at the Dominguez Field (renamed Aviation Camp). He set a new altitude record of 4,165 feet. Located at what is now Wilmington Avenue south of Artesia Boulevard, Aviation Camp was described as "a mesa land [that] slopes to the north [with] conditions favorable for a turn to the west and south, and a sweep over the Dominguez Dairy and the lowlands west of Willows." Aviators flew so close to Long Beach that the *Long Beach Press* warned readers that "aero planes may take a notion to sail over Long Beach for a time." Four days before the event, the *Long Beach Press* published "Rules for Airships," cautioning the "queer contraptions" not to curl around "the tall buildings of Long Beach" or to "run races with street cars here. No alighting on auditorium flagpoles or minarets. Seek a soft spot out near Belmont Heights for landing purposes. Aviators catching cows in their dragropes will please return them to owners within a reasonable time. Please don't allow the cows to strain their milk." (LOC.)

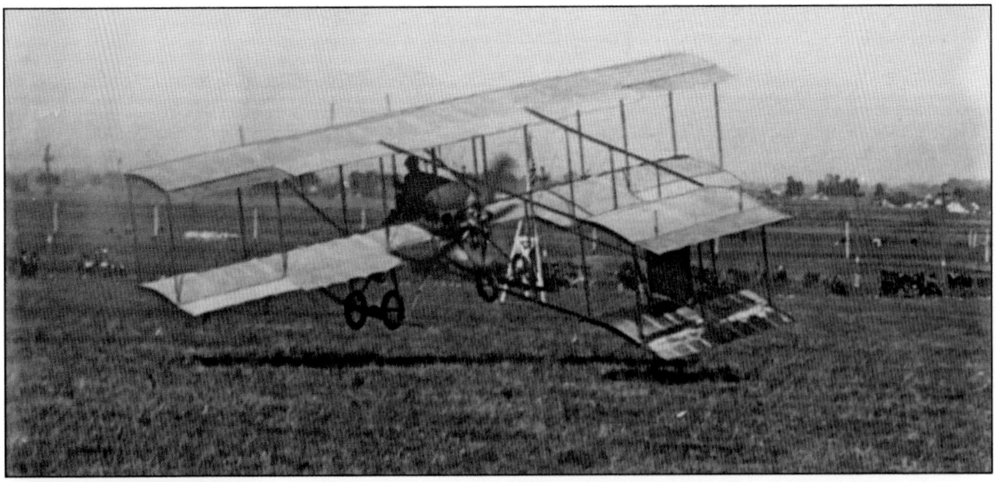

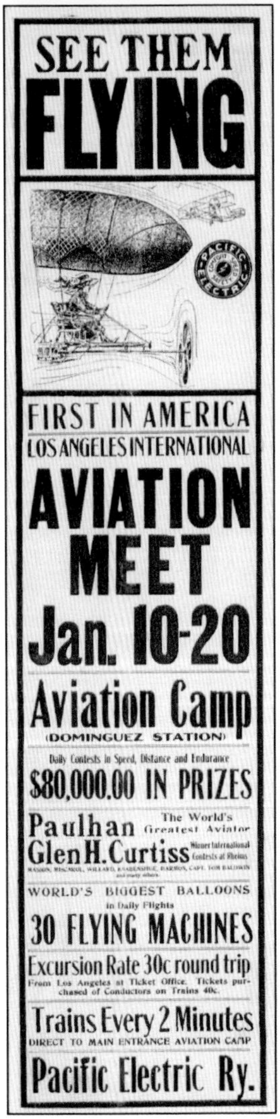

International Air Meet. Ticket sales topped 254,000 for the January 10–20, 1910, Los Angeles International Air Meet as people came by bicycle, horseback, automobile, and "every two minutes by trolley" to see the skies filled with all types of flying contraptions, with "birdmen" performing daring flights and maneuvers. The *Los Angeles Times* reported it was "one of the greatest public events in the history of the West." William Randolph Hearst took his first flight in Louis Paulhan's airplane (pictured) and a year later offered $50,000 for the first transcontinental flight, which would land in Long Beach. (Above, LOC; left, CSUDH.)

14 SECOND ANNUAL AVIATION MEET

Programme

(Continued from page 13)

Tuesday, December 27

"PASADENA DAY"

(Under the Regulations of the International Aeronautical Federation)

HOURLY CONTESTS

World's Records	Daily Speed
Daily Altitude	Accuracy in Landing
Quick Start	Bomb Throwing
Passenger Carrying	Perkins' Kites

The beginning and ending of Events will be signalled by a bomb or cannon. Events, results and progress of Events will be announced on the blackboard. See Code and Signals, pages 20 to 25.

Wednesday, December 28

"LONG BEACH DAY"

(Under the Regulations of the International Aeronautical Federation)

HOURLY CONTESTS

World's Records	Daily Speed
Daily Altitude	Accuracy in Landing
Quick Start	Bomb Throwing
Passenger Carrying	Perkins' Kites

The beginning and ending of Events will be signalled by a bomb or cannon. Events, results and progress of Events will be announced on the blackboard. See Code and Signals, pages 20 to 25.

(Continued on page 15)

OFFICIAL PROGRAMME 15

Programme

(Continued from page 14)

Thursday, December 29

"ORANGE BELT DAY"

(Under the Regulations of the International Aeronautical Federation)

HOURLY CONTESTS

World's Records	Daily Speed
Daily Altitude	Accuracy in Landing
Quick Start	Bomb Throwing
Passenger Carrying	Perkins' Kites

The beginning and ending of Events will be signalled by a bomb or cannon. Events, results and progress of Events will be announced on the blackboard. See Code and Signals, pages 20 to 25.

(Continued on page 16)

Music by
PORTER'S CATALINA ISLAND BAND
40 Pieces
W. F. Arend, Director

LONG BEACH TROPHY. December 28, 1910, was Long Beach Day at the Second Los Angeles International Air Meet, held December 24, 1910–January 3, 1911. The chamber of commerce sponsored the Long Beach Trophy for altitude. Arch Hoxsey, 26, won, setting an 11,474-foot record. Determined to break his own record, Hoxsey tried on December 31, 1910, to reach 12,000 feet, ignoring the warnings from other aviators of strong "gale winds" in the area. He went up in the bad weather and spun out of control around 7,000 feet. Hoxsey's Long Beach Trophy was displayed posthumously in the window of a Long Beach jewelry shop on Pine Avenue. Two months prior to his death, Hoxsey gave ex-president Theodore Roosevelt his first ride in an airplane. Upon landing, Roosevelt remarked, "That was the bulliest experience I ever had." (Above, CSUDH; below, LOC.)

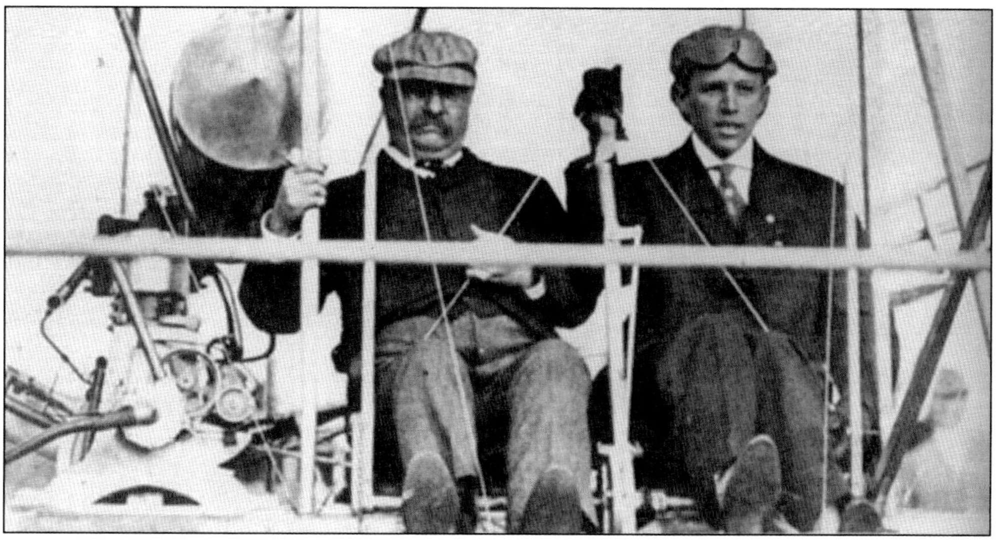

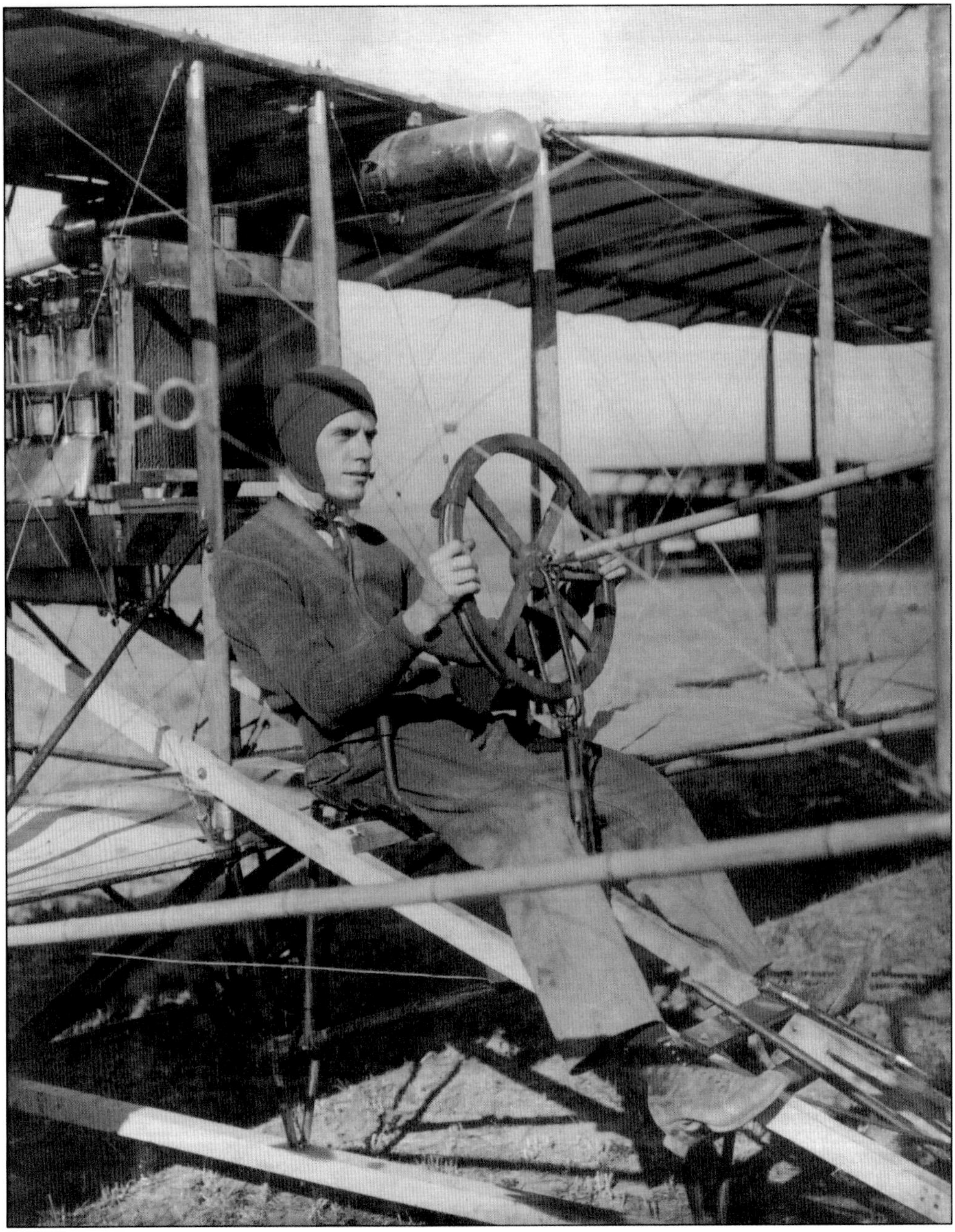

First Long Beach Flyers. The Dominguez event was significant in the development and growth of aviation in the nation and specifically in Long Beach. William Boeing traveled from Seattle, Washington, in hopes of getting his first ride in an airplane but failed to find someone willing to fly him. Eventually Boeing established a major airplane manufacturing company that would partner with Douglas Aircraft, a major Long Beach employer, during World War II. The Los Angeles air meets attracted several Long Beach residents: Frank Champion, J. A. Boyton, Frank Stites, and Earl Daugherty. Daugherty (pictured) is the aviator credited for establishing the first inland airfield in Long Beach and for encouraging the city council to establish its municipal airport in 1923.

Three

Earl S. Daugherty Gets His Start in Aviation

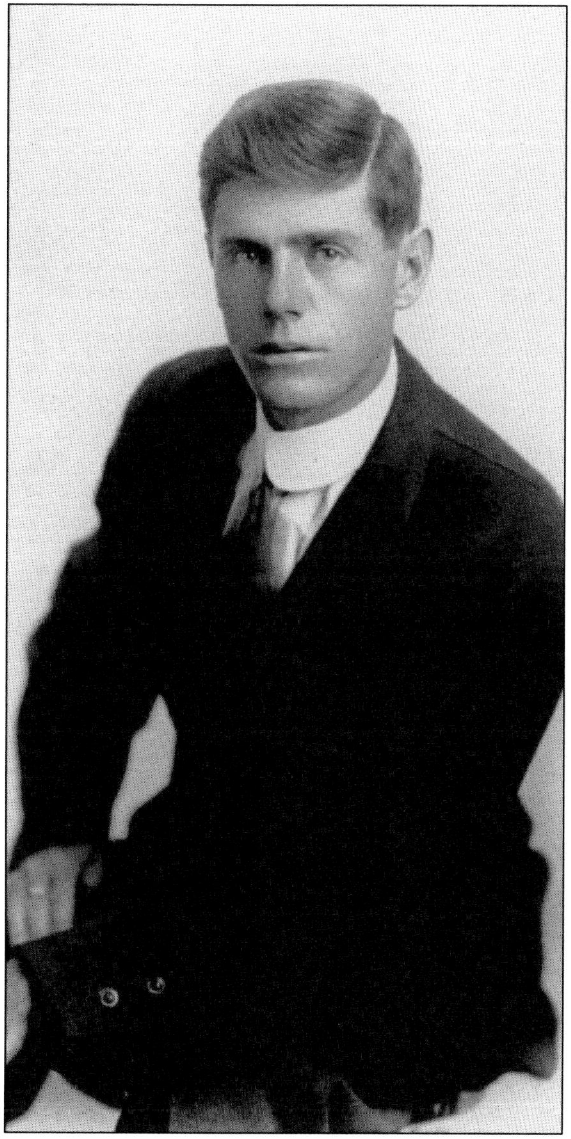

Earl S. Daugherty. Earl S. Daugherty attended the 1910 Los Angeles Air Meet with his two friends Frank Champion and J. A. Boyton. The three helped the airplanes take flight by holding onto the wings and giving a push until the flyers gave the signal "to let go."

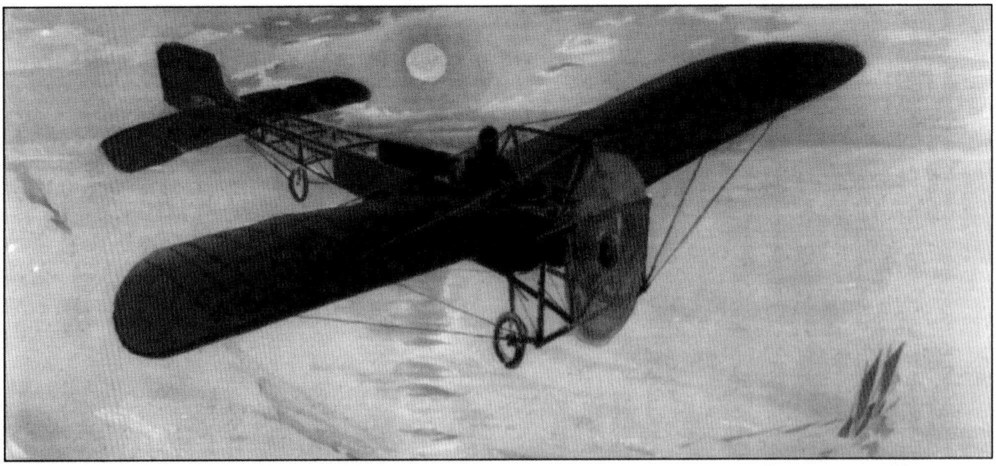

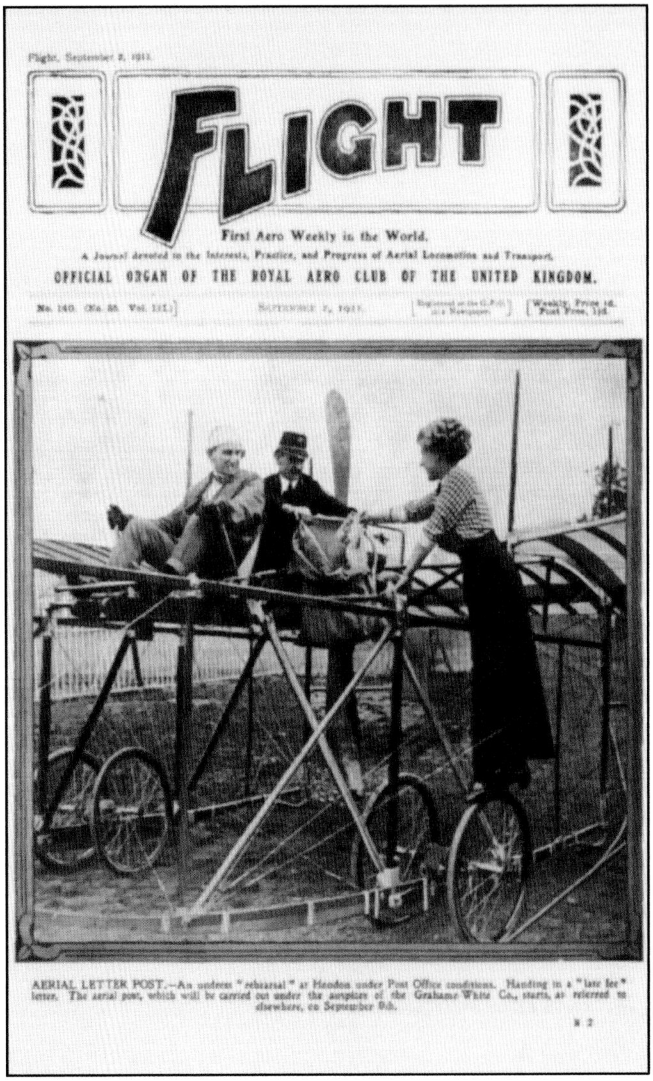

Flight, September 2, 1911.

FLIGHT

First Aero Weekly in the World.

A Journal devoted to the Interests, Practice, and Progress of Aerial Locomotion and Transport.

OFFICIAL ORGAN OF THE ROYAL AERO CLUB OF THE UNITED KINGDOM.

No. 140. (No. 35. Vol. III.) SEPTEMBER 2, 1911. [Registered at the G.P.O. as a Newspaper.] [Weekly, Price 1d. Post Free, 1½d.]

AERIAL LETTER POST.—An undress "rehearsal" at Hendon under Post Office conditions. Handing in a "late fee" letter. The aerial post, which will be carried out under the auspices of the Grahame-White Co., starts, as referred to elsewhere, on September 9th.

B 2

FRANK CHAMPION. Champion moved to Long Beach after a stint in the navy and operated a photography concession at the Pike called Burro Studios. After the 1910 air meet, he enrolled in the Bleriot Aviation School at Hendon, England. Bleriot developed the single wing (monoplane) airplane (pictured above) that Champion had watched Louis Paulhan fly in and win the $10,000 grand prize at the Dominguez Air Meet. Paulhan flew the airplane across the English Channel in 1909. Champion became the first person in Long Beach (and the 86th in the nation) to get a license from the Aero Club of America (an affiliate of the Automobile Club of America), which sanctioned competitive flying events. *Flight Magazine* (official publication of the United Kingdom Royal Aero Club) on September 2, 1911, reported Frank L. Champion's return to Long Beach and his engagement "to fly at the Long Beach Carnival Festivities." (Above, LOC; left, www.flightglobal.com.)

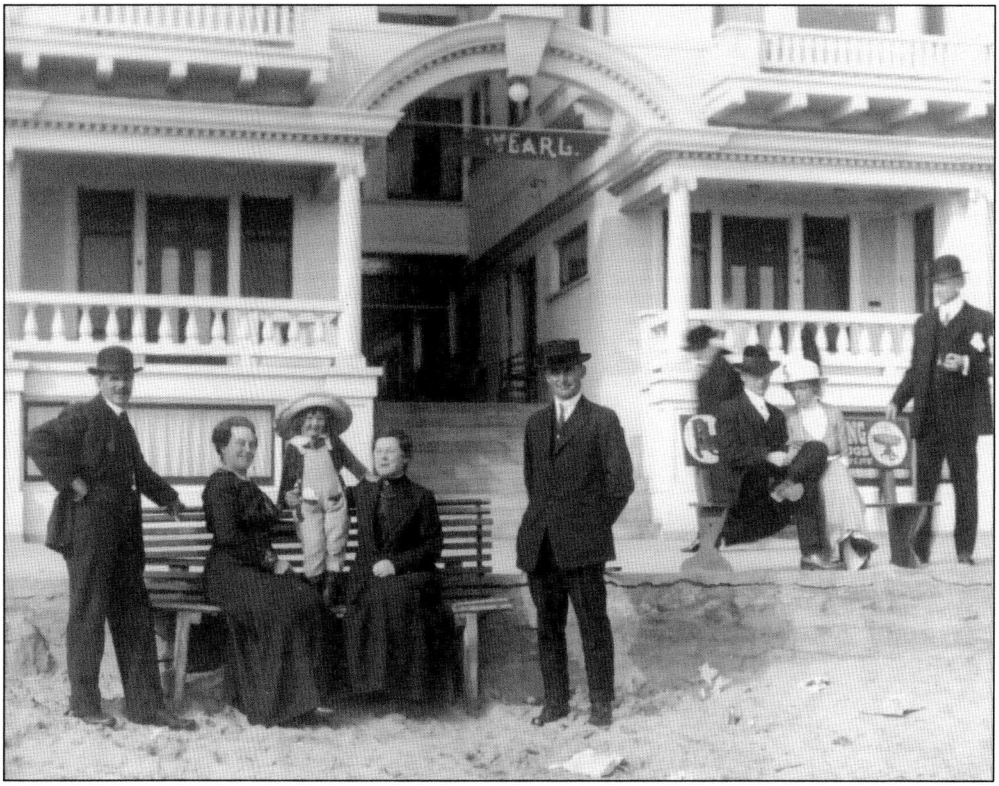

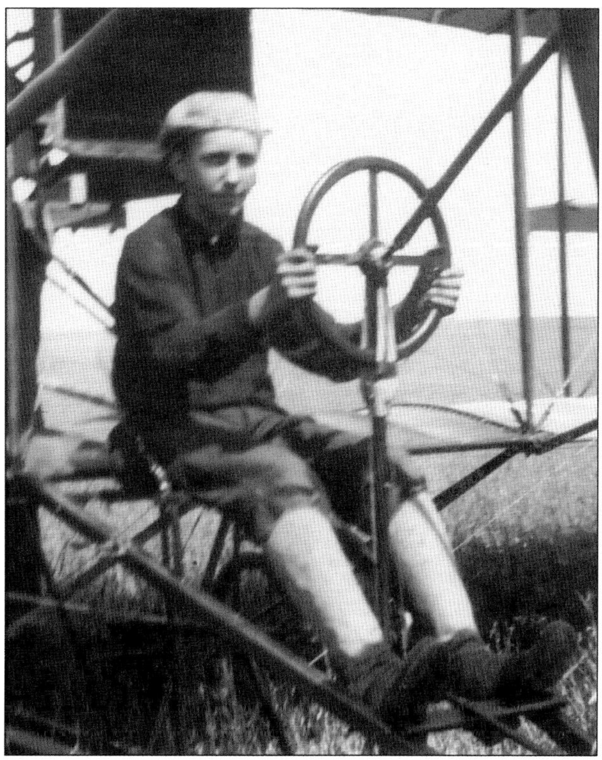

The Earl. Earl Daugherty (right), born 1887 in Des Moines, Iowa, moved with his parents, Charles and Mary, to Long Beach in 1902. Daugherty's father was a photographer and a real estate investor. Daugherty attended Long Beach schools, worked briefly at a bank, and was a member of the El Rodeo Club along with the sons of prominent families, including the Bixby land barons. Daugherty's parents owned the Earl apartments (named after him), which sat on an incline just steps from the Pike on the beach. He and Champion launched the airplanes they built together in the basement of the Earl. Together they would also have a national career of testing and flying airplanes. Champion, dubbed the "Candy Kid" in aviation magazines, earned a good living charging as much as $1,500 for a 15-minute flight in his Bleriot plane.

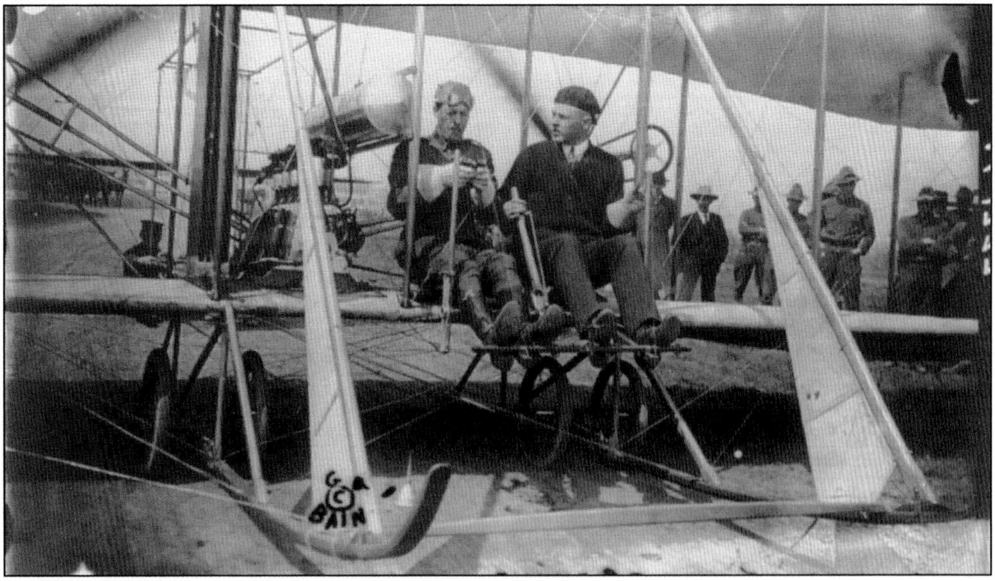

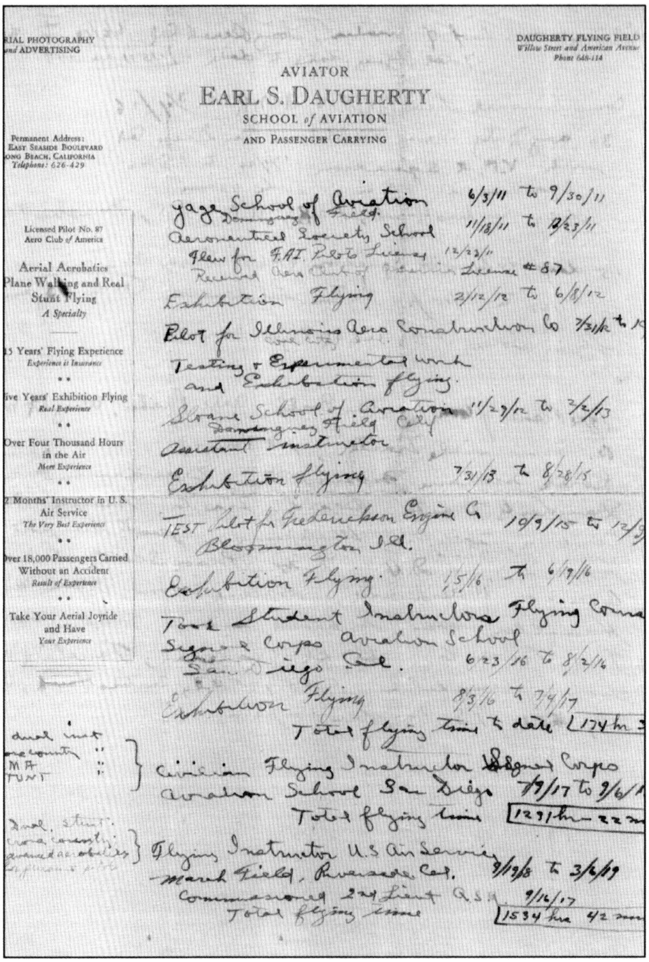

RIAL PHOTOGRAPHY
and ADVERTISING

DAUGHERTY FLYING FIELD
Willow Street and American Avenue
Phone 648-114

AVIATOR
EARL S. DAUGHERTY
SCHOOL of AVIATION
AND PASSENGER CARRYING

Permanent Address:
East Seaside Boulevard
Long Beach, California
Telephone: 626-429

Licensed Pilot No. 87
Aero Club of America

Aerial Acrobatics
Plane Walking and Real
Stunt Flying
A Specialty

15 Years' Flying Experience
Experience is Insurance

Five Years' Exhibition Flying
Real Experience

Over Four Thousand Hours
in the Air
More Experience

2 Months' Instructor in U.S.
Air Service
The Very Best Experience

Over 18,000 Passengers Carried
Without an Accident
Result of Experience

Take Your Aerial Joyride
and Have
Your Experience

Gage School of Aviation 6/3/11 to 9/30/11
Dominguez Field
Aeronautical Society School 11/18/11 to 12/23/11
Flew for F.A.I. Pilots License 12/23/11
Exhibition Flying 2/12/12 to 6/8/12
Pilot for Illinois Aero Construction Co
Testing & Experimental work
and Exhibition flying
Sloane School of Aviation 11/29/12 to 3/2/13
Dominguez Field Calif
Assistant instructor
Exhibition flying 7/31/13 to 8/28/15
TEST Pilot for Frederickson Engine Co 10/9/15 to
Bloomington Ill.
Exhibition Flying 6/9/16
Took Student Instructors Flying Course
Signal Corps Aviation School
San Diego Cal. 6/23/16 to 8/2/16
Exhibition Flying 8/3/16 to 7/4/17
Total flying time to date 174 hrs
Civilian Flying Instructor Signal Corps
Aviation School San Diego
Total flying time 1291 hrs 22 min
Flying Instructor U.S. Air Service
March Field, Riverside Cal. to 3/6/19
Commissioned 2nd Lieut A.S. 9/16/17
Total flying time 1534 hrs 42 min

FLYING SCHOOLS. Four schools of aviation operated near Dominguez Field by 1911. Other flying schools included the following: Wright School was run with Phil Parmalee (pictured above with Lt. Benjamin Foulois) and Clifford Turpin as instructors; Glenn Martin of Santa Ana started a school at Griffith Park; Glenn Curtiss taught students on Coronado Island near San Diego; the Queen Monoplane Company offered lessons in Pasadena; and balloonist Capt. Thomas Baldwin operated a school in Los Angeles. Daugherty's training started with flying lessons from the Gage School of Aviation at Dominguez Field from June 3, 1911, to September, 30, 1911. He also took lessons from the Aeronautical School from November 18, 1911, to December 23, 1911, as indicated in the notes he jotted on his personal stationery. For the most part, Daugherty and early aviators taught themselves to fly. (Above, LOC.)

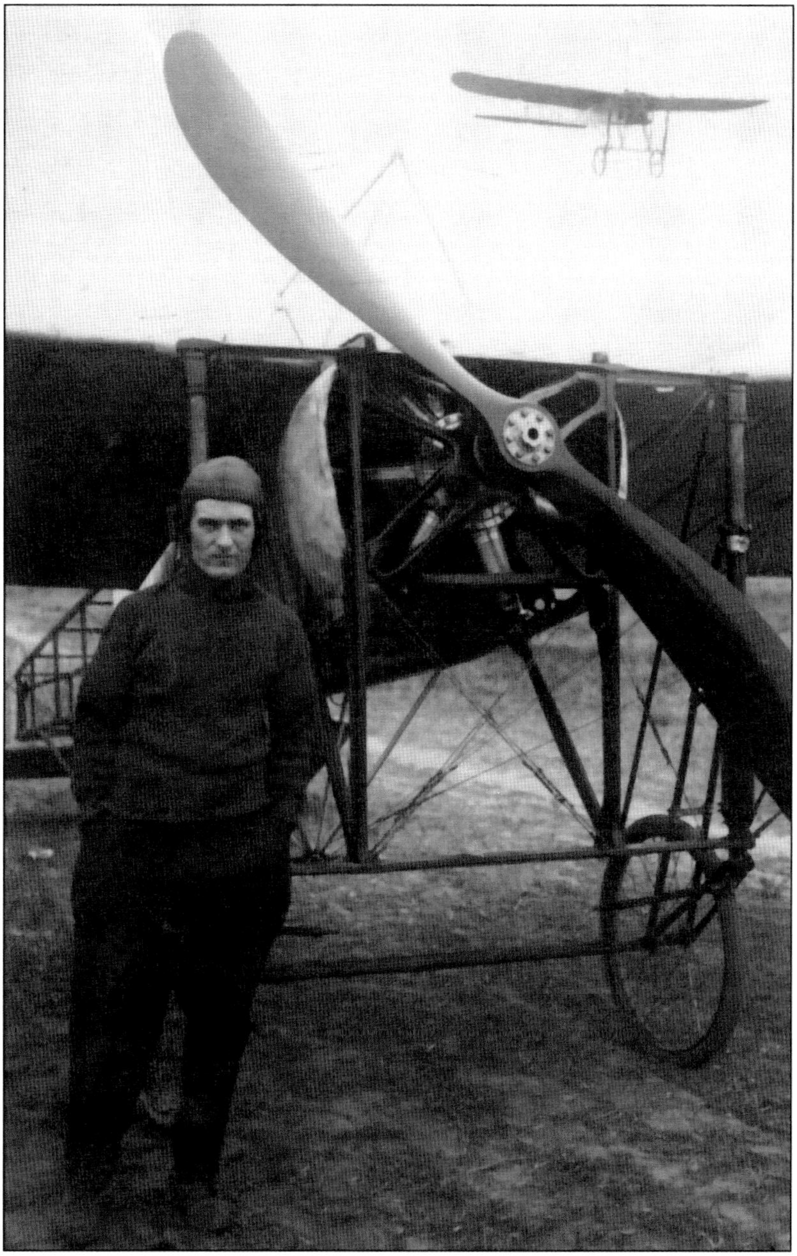

Long Beach Firsts. *AERO Magazine* reported that Saturday, December 23, 1911, at Dominguez Field, Los Angeles, California, "Frank Stites, Frank Champion [pictured], Earl Dougherty [*sic*] and Harry Holmes qualified for pilot's certificates. Stites used a biplane of his own make with Macomber rotary motor; Holmes and Dougherty [*sic*] used Walsh's old Curtiss-type with Hall-Scott motor; Champion used a Bleriot-Gnome. Beryl Williams also flew." Daugherty's friend Frank Champion flew from Dominguez Field to Long Beach to drop a note to Daugherty's parents: "Dear Father and Mother, I made my first successful 15 minute flight in the Curtiss biplane. Will send this by Frank Champion in his Bleriot. Your son, Earl." Daugherty carried his first passenger, Nell Shall, during his qualifying flight. The success of these aviators gave Long Beach the largest number of internationally licensed pilots in the West in 1911.

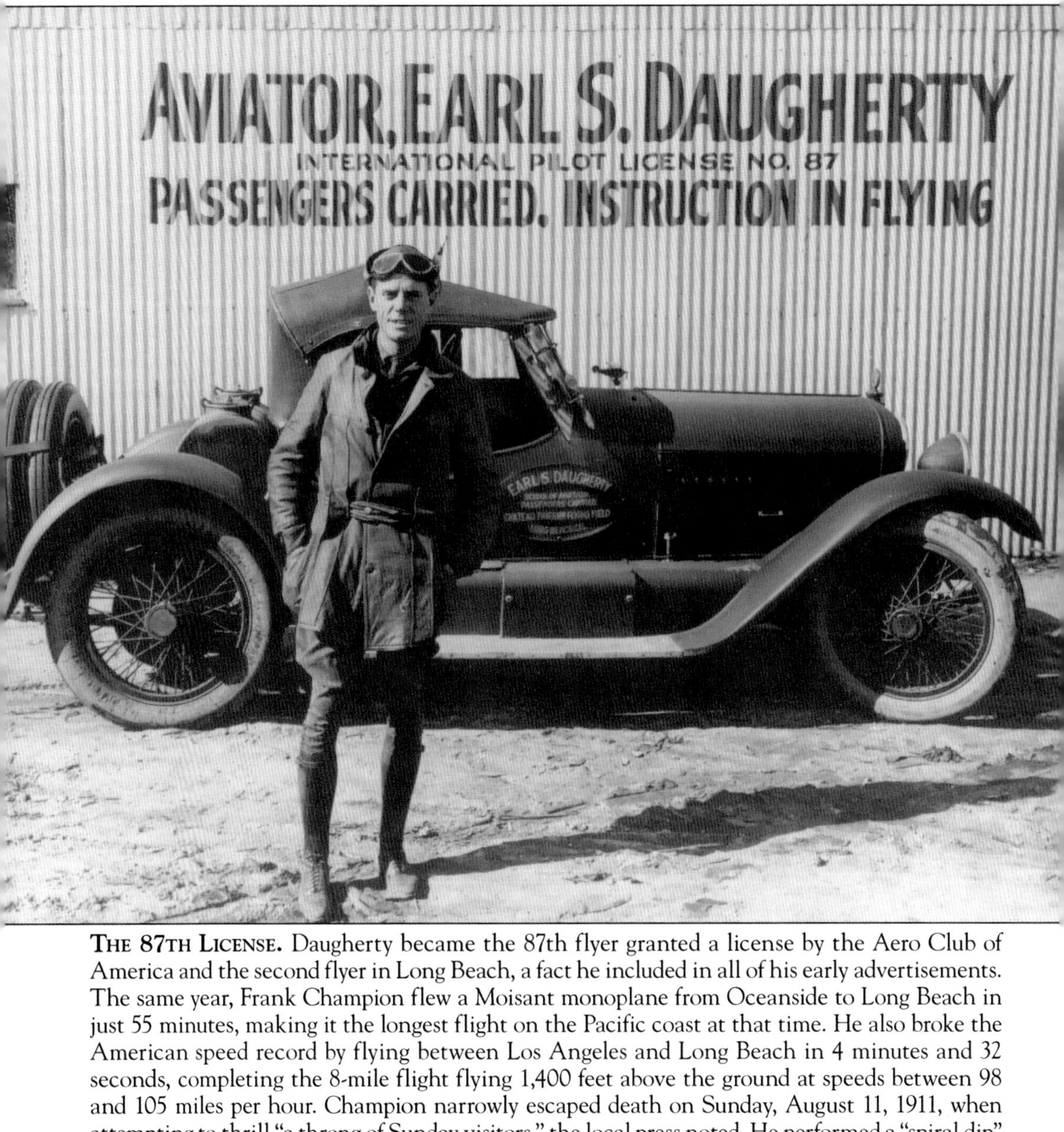

THE 87TH LICENSE. Daugherty became the 87th flyer granted a license by the Aero Club of America and the second flyer in Long Beach, a fact he included in all of his early advertisements. The same year, Frank Champion flew a Moisant monoplane from Oceanside to Long Beach in just 55 minutes, making it the longest flight on the Pacific coast at that time. He also broke the American speed record by flying between Los Angeles and Long Beach in 4 minutes and 32 seconds, completing the 8-mile flight flying 1,400 feet above the ground at speeds between 98 and 105 miles per hour. Champion narrowly escaped death on Sunday, August 11, 1911, when attempting to thrill "a throng of Sunday visitors," the local press noted. He performed a "spiral dip" and plunged into the surf 100 feet from the shore. Bathers and lifeguards swam to his assistance and unfastened the straps that held him in his seat.

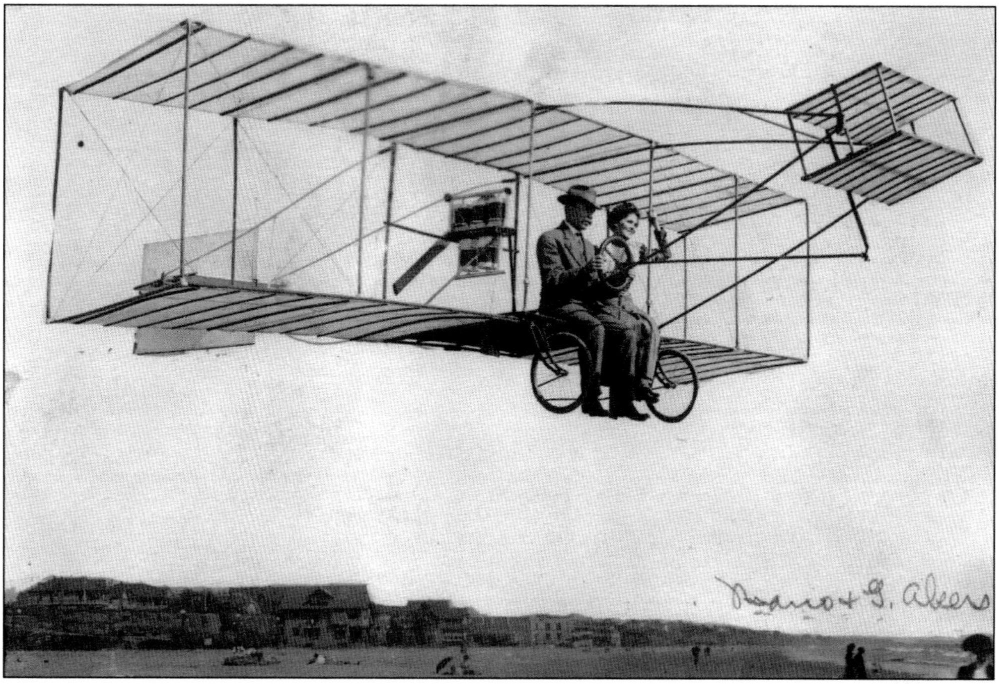

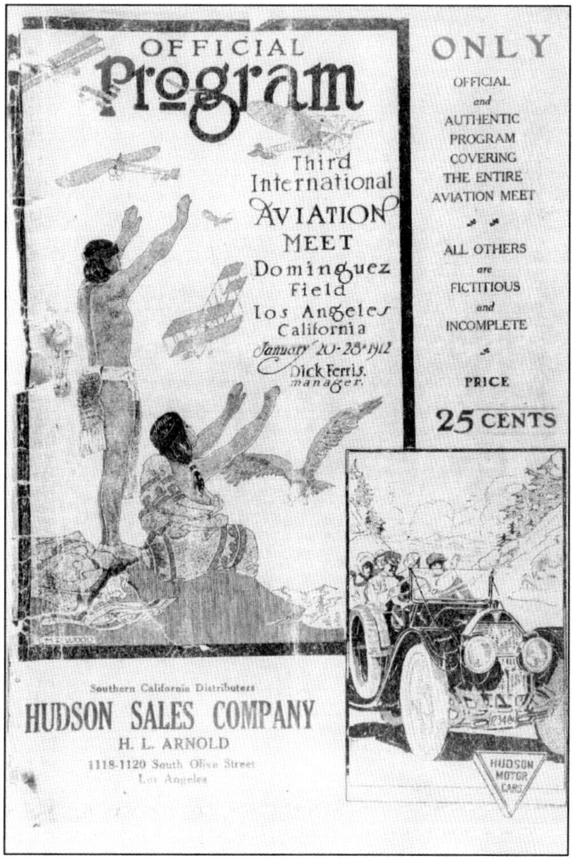

LONG BEACH FLYING MACHINES. Through the efforts of Dr. Harriman Jones and William P. Nestle, manager of the Hotel Virginia, Frank Champion and Earl Daugherty were able to build planes in the basement of the hotel and entertain crowds as they flew from the beach. Visitors had photographs taken in a model airplane that was superimposed over the backdrop of downtown Long Beach (above). Together and separately, Champion and Daugherty crisscrossed the country earning a living as stunt flyers and test pilots, trying out the designs of the latest airplanes. Champion worked with Earle Remington, who took over operations at Dominguez Field and helped set up the third and final Los Angeles Air Meet in 1912. Remington manufactured planes at his California Aviation Company. (Right, CSUDH.)

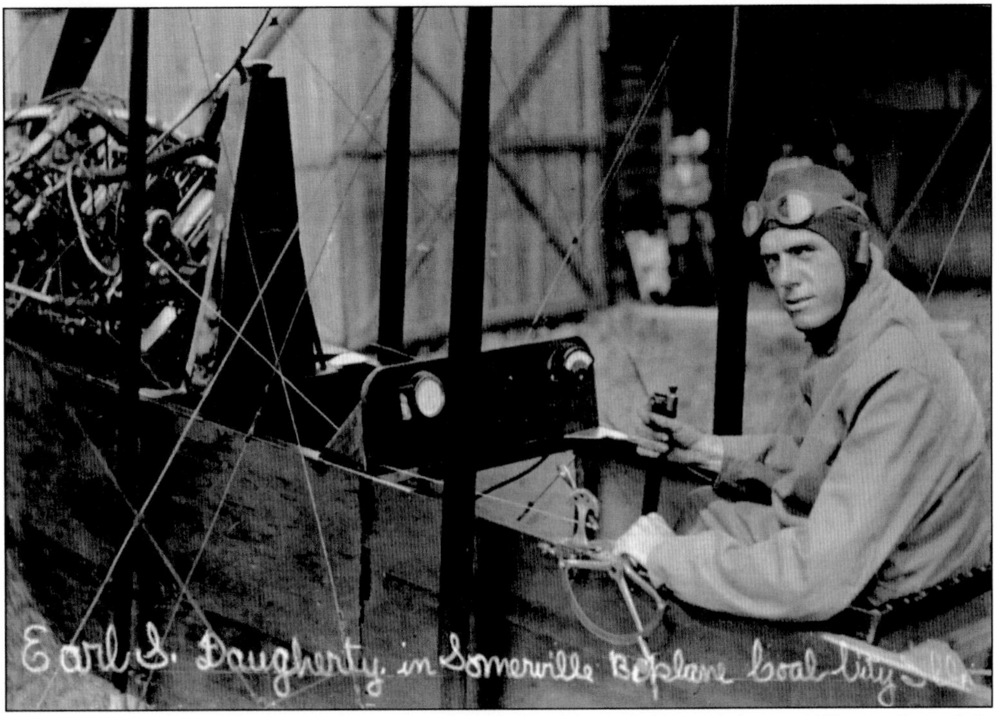

Illinois Aviation. Daugherty's flying and plane-building skills took him to Illinois from July 31, 1912, to October 26, 1912, where he did testing, experimental work and exhibition flying for the newly formed Illinois Aero Construction Company in Coal City. Local newspapers reported his racing a biplane against two passenger trains and "playing tag with another at the Grundy county fair" to the delight of his passenger, Kate Miller (a telephone operator) of Coal City. That year, Frank Champion was shot accidentally and almost lost his foot to gangrene. Long Beach aviators held a benefit ball for him. Eventually he had leg surgery but had problems "shaking the voodoo" of his leg problem. In June 1912, Champion thrilled thousands on the beach in Long Beach with his stunts. While Daugherty was in Illinois, Champion went to Overland, Kansas, to start the Frank Champion Aeroplane Company Factory and School. He moved his company to Chicago in 1916.

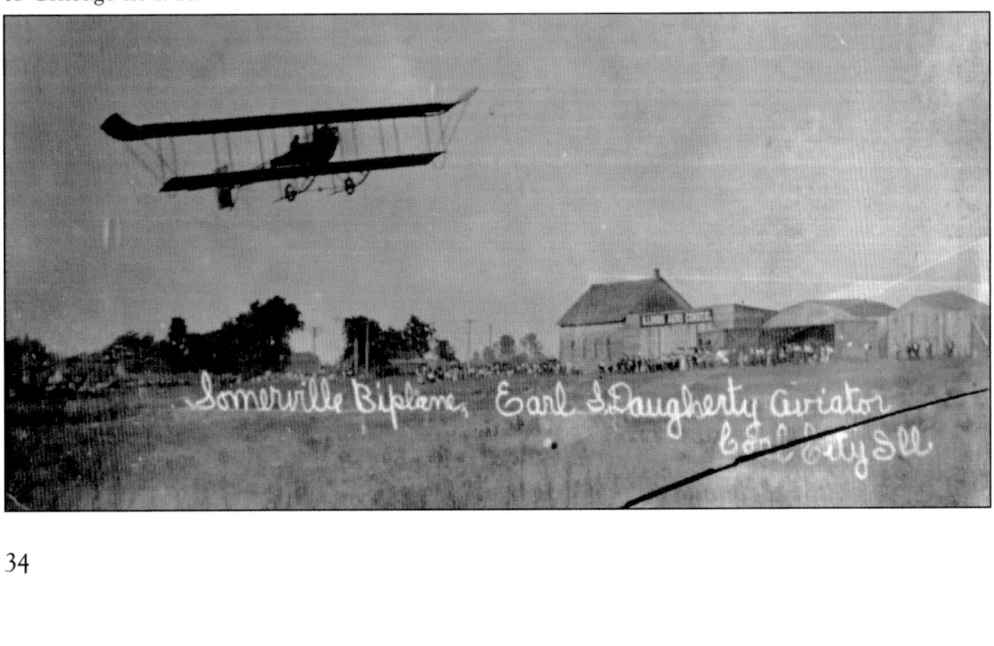

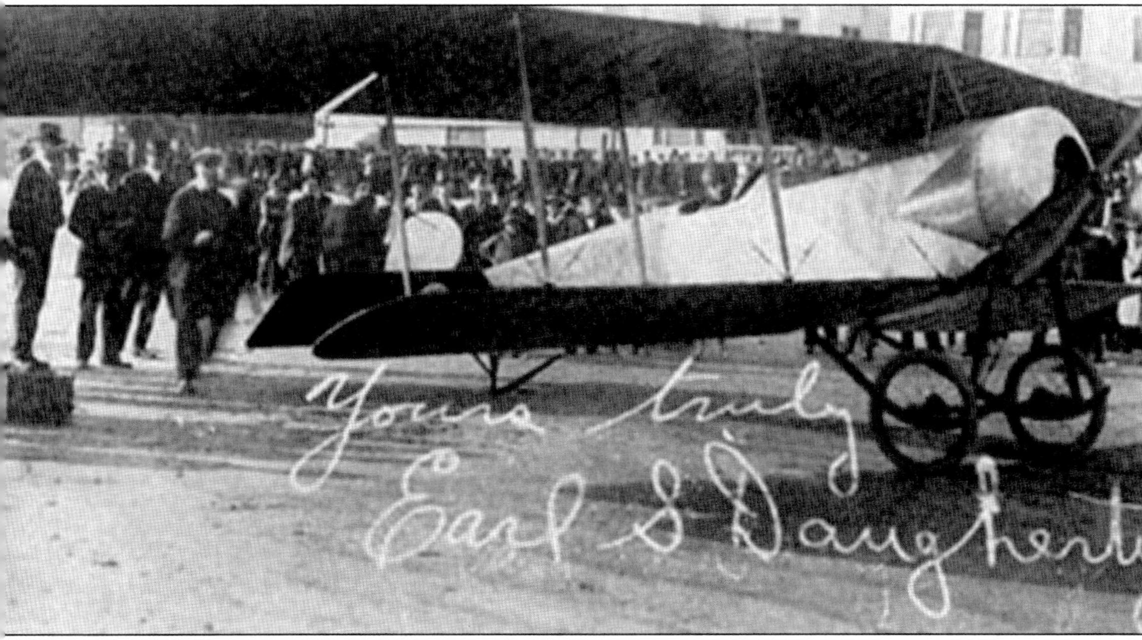

Stable Biplane. In September 1912, Daugherty flew an "Illinois Aero Construction Company Auto Stable Biplane" in the Cicero-Aurora Aviation Meet, making airmail drops with fellow aviator Max Lillie and competing with a group of aviators in a "looping and mock bomb dropping on mock battleships" contest. The contest was important to the military, which was examining the possibilities of using airplanes in war. Daugherty narrowly escaped death when he and a passenger plunged over a half mile to the earth. Daugherty returned to the West Coast and competed in the 1912 Dominguez Air Meet, winning the title of one of the "Promising Los Angeles Flyers." Daugherty kept a hangar at the Cicero Field in Illinois. He had another close call when in December 1912 the propeller of his Deperdussin flew to pieces and sent the engine flying off in another direction. The same year, Daugherty filled in for Frank Stites, who broke his arm and could not perform stunts for a movie being filmed in California.

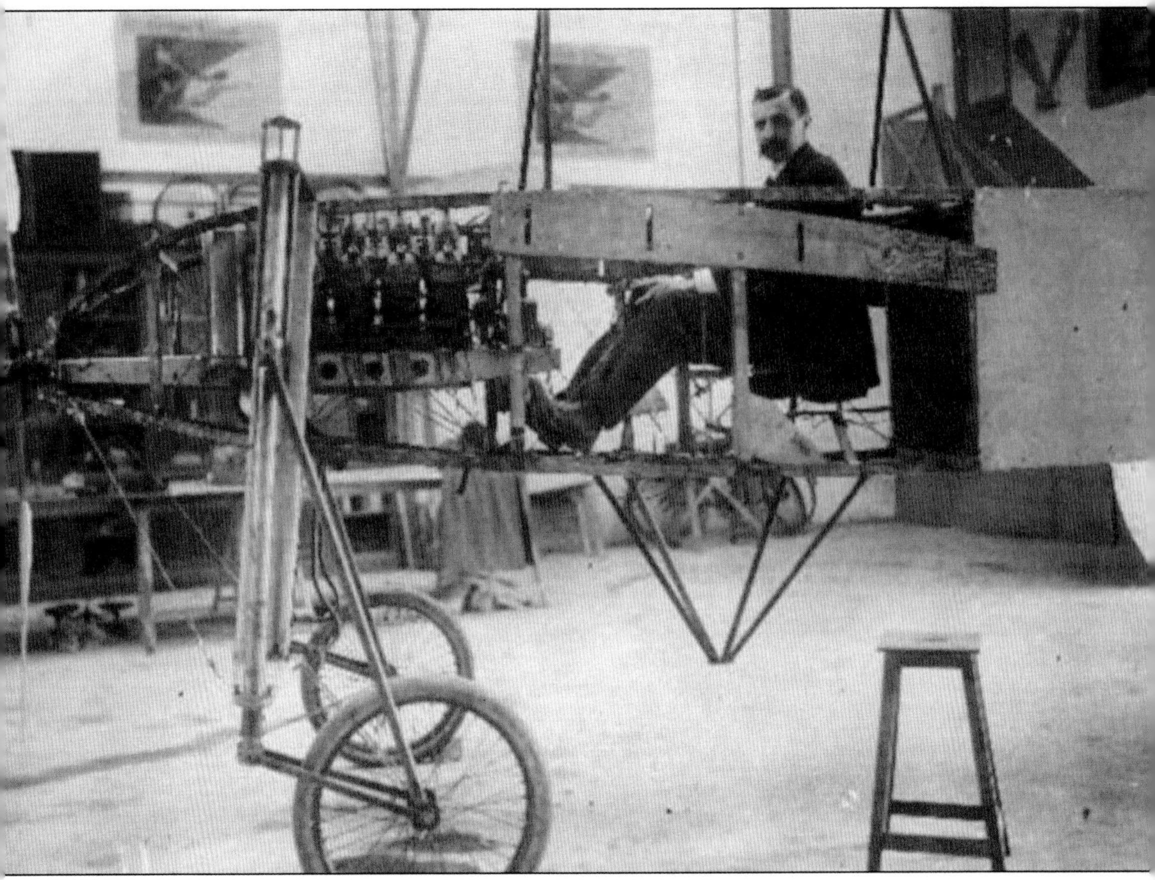

Bleriot Plane. In 1913, Frank Champion shipped by train to Cicero Field the Bleriot XI-type plane he and Daugherty built. (This photograph shows Louis Bleriot in his XI plane.) Daugherty and Champion also signed on with the Young Aeroplane Company in Waterloo, Iowa. The *Waterloo Times-Tribune* explained, "These aviators are well known throughout the United States and a large number of them will be in Waterloo a part of the time next summer. They have been known as Young's High Flyers from Kansas City but will hereafter be booked from Waterloo. It is also planned to have two or three of these aviators make a trip from Waterloo to the coast this winter. This matter has not been fully decided upon as yet, but Mr. Young stated recently that he was quite sure the men would make the trip. This would be a great advertising feature for this city and do much toward placing this city on the aero plane map." (LOC.)

LONG BEACH FLYER. Champion and Daugherty continued flying and building airplanes together. Designed by Daugherty and built by Max Stupar, the "Daugherty Stupar Tractor," dubbed the *Long Beach Flyer*, was the first plane used to advertise a city. Daugherty's design was featured in the November 14, 1914, edition of *Aero and Hydro Magazine*. Airplane design changed quickly, as can be seen in the photograph of Daugherty sitting on the right on a wing of an early plane with his friend and fellow military flight instructor, John Montijo, in the pilot's seat. The other man is unidentified. Daugherty's "Stupar Tractor" was aerodynamically different from earlier planes and powered by an engine in the front, which pulled the plane instead of pushing it, making it suitable for military or exhibition flying.

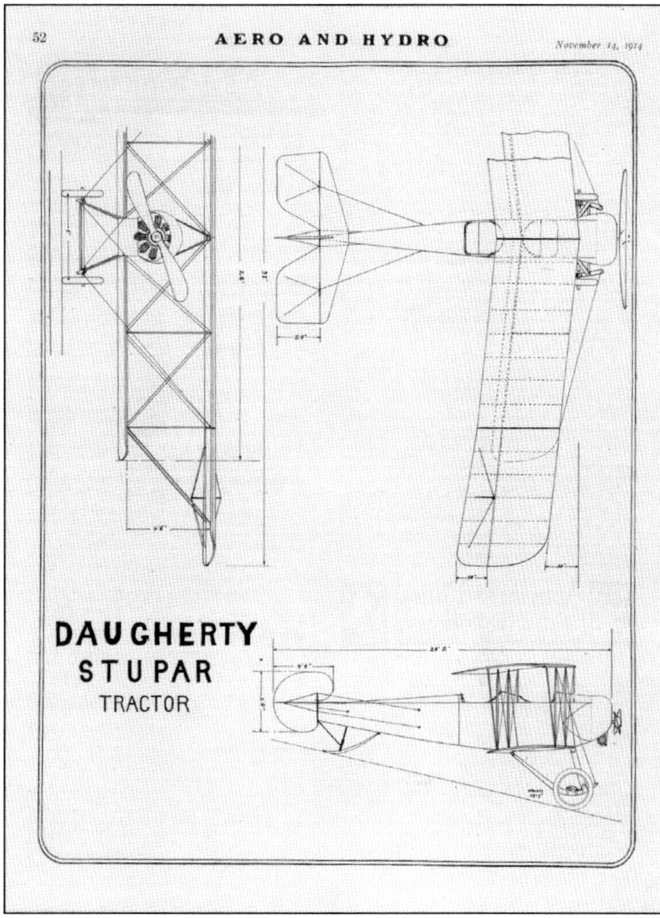
52 AERO AND HYDRO November 14, 1914

DAUGHERTY
STUPAR
TRACTOR

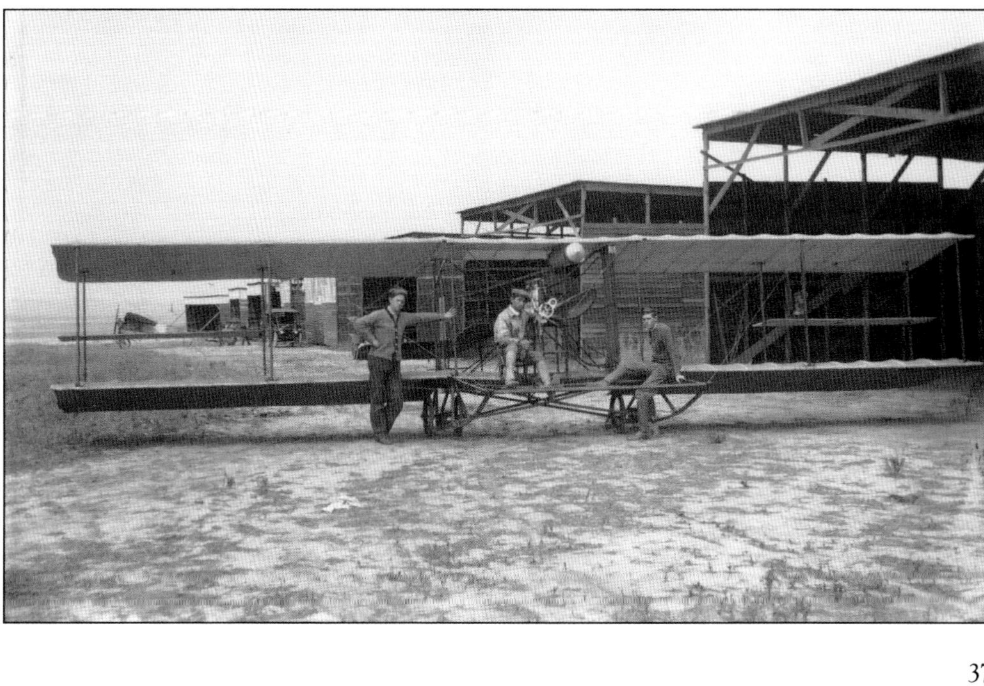

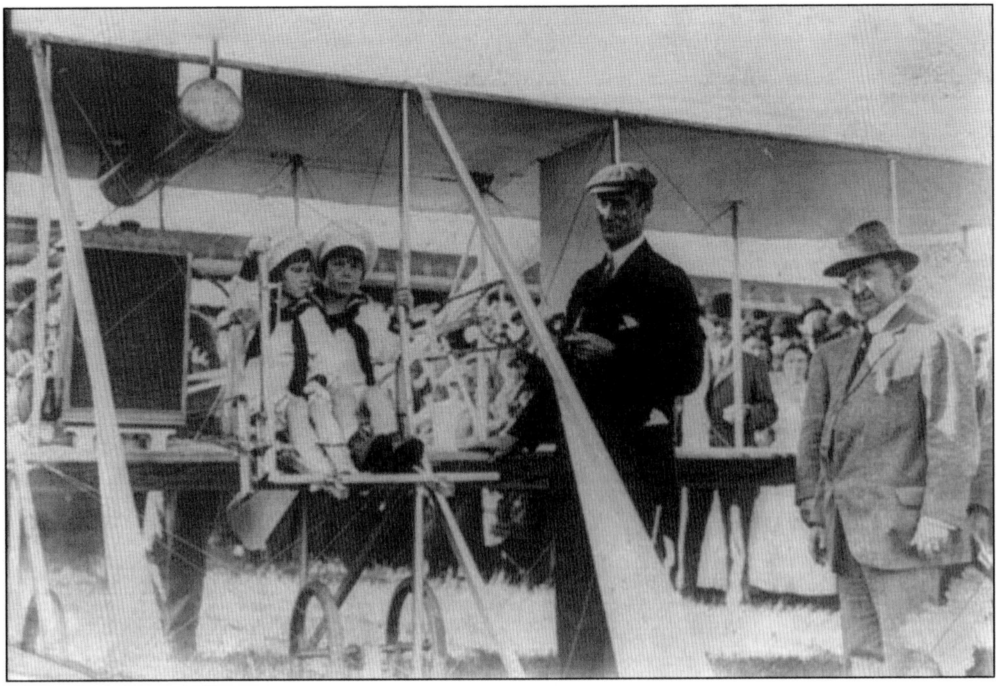

Cal Rodgers. Other famous aviators chose Long Beach as the site on which to make their mark. On December 10, 1911, Calbraith Perry ("Cal") Rodgers (pictured with trademark cigar) completed the first U.S. transcontinental flight from Sheepshead Bay, New York, in the surf off Long Beach. William Randolph Hearst offered a $50,000 prize to the first aviator completing a flight coast-to-coast within 30 days. Rodgers, the son of naval hero Oliver Perry, convinced the Armour Company to finance his race, and in turn, he displayed a "Vin Fiz" advertisement (Armour's new grape soft drink) on the wing of the Wright Model EX-1 pusher plane. (Both, LOC.)

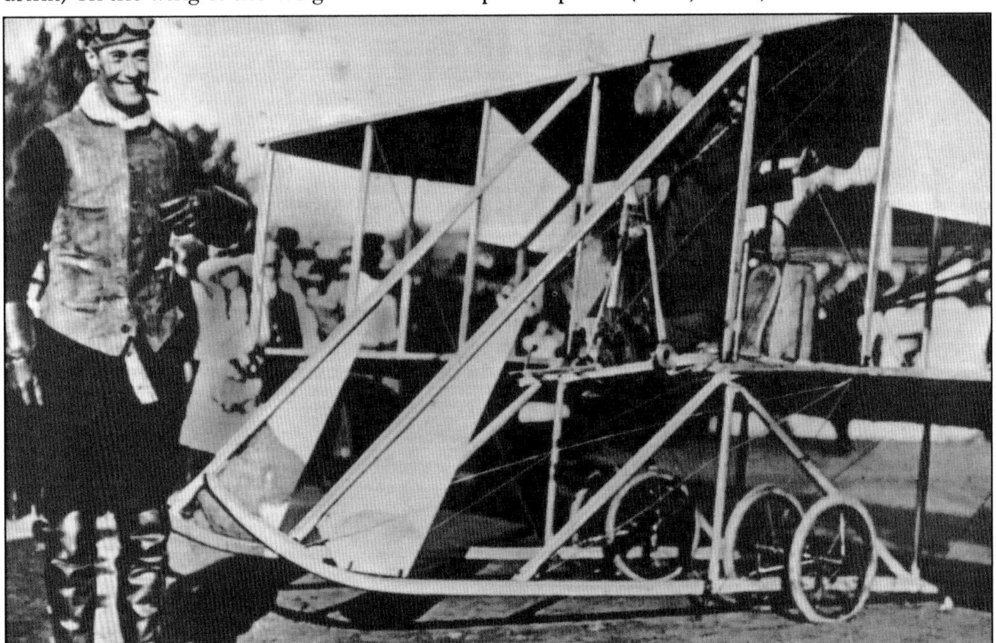

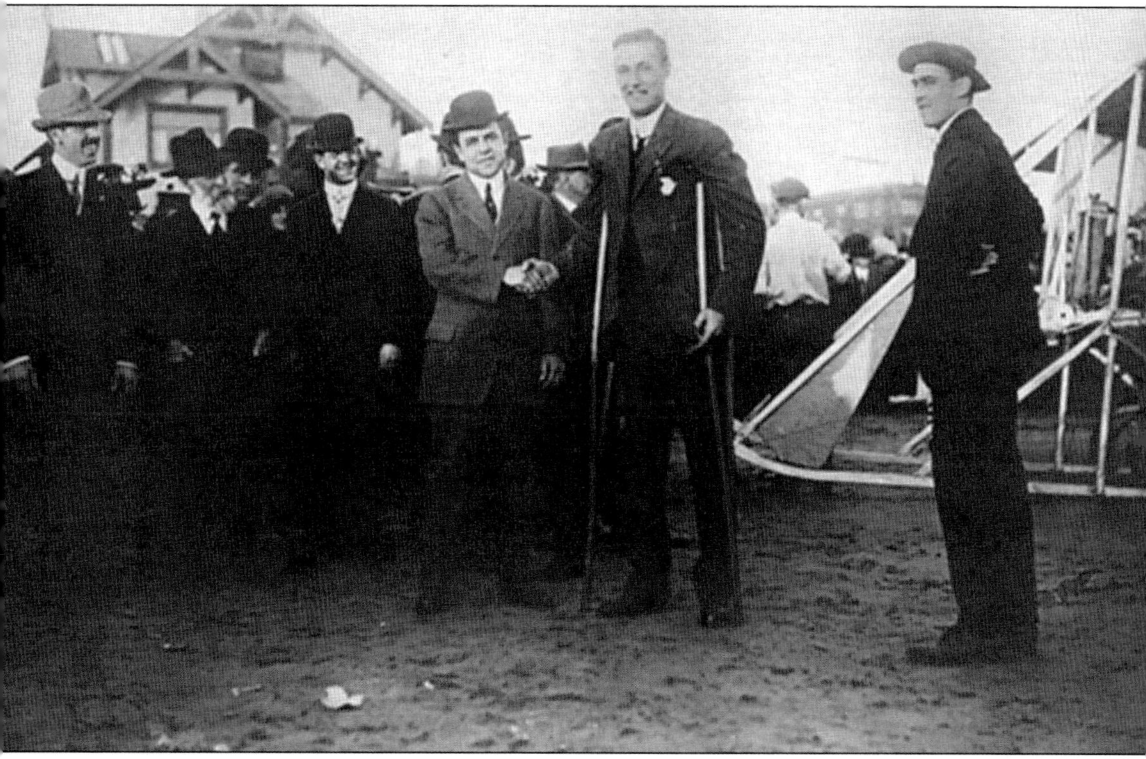

Landing in Long Beach. The chamber of commerce enticed Rodgers with $1,000 to end his trip in Long Beach instead of Pasadena. Rodgers landed in Pasadena, resumed his flight, and crashed again before landing in the surf off Long Beach. Rodgers missed Hearst's deadline and lost the prize because of delays caused by numerous crashes. More than 50,000 people lined up to watch Rodgers drop into the surf as his friends Frank Champion in a Bleriot monoplane and Beryl "Byrd" Williams and Earl Daugherty (who would qualify for his license later) in a Curtiss-type plane flew out to meet Rodgers as he passed Dominguez Field and give him an escort into Long Beach. All three airplanes circled the water and landed on the sand and water near the Pine Avenue Pier. Rodgers used crutches when he landed in Long Beach because of his numerous injuries. A replica of the *Vin Fiz* (because little remained of the original) is on display at the Smithsonian Air and Space Museum in Washington, D.C., because of its significance in aviation history.

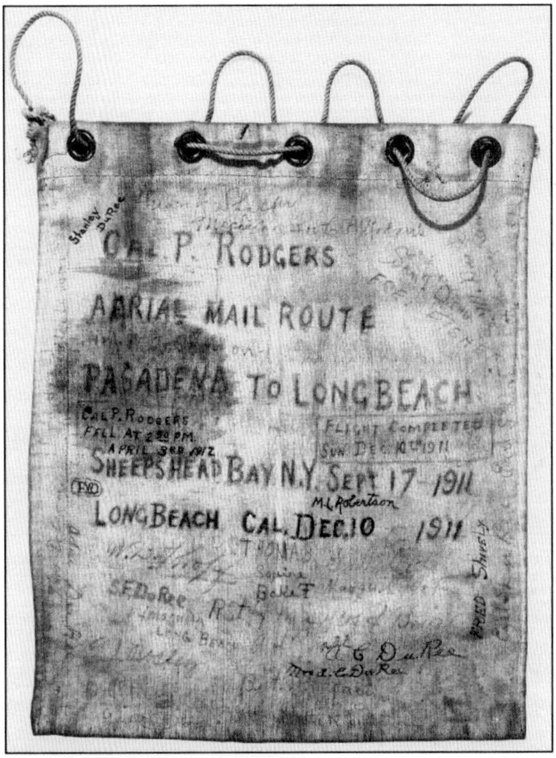

Aerial Mail. Rodgers carried an "aerial mail" bag from Pasadena that he dropped in Long Beach before landing. The bag was signed by local aviation supporters, such as Squire DuRee, Frank Shaffer, and Earl S. Daugherty.

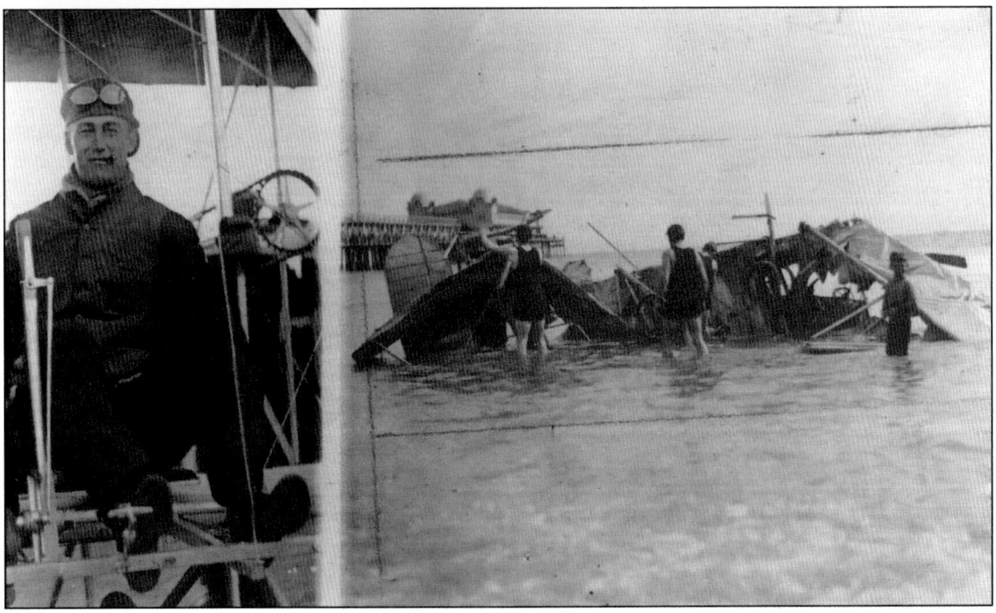

Dangerous Birds. On April 3, 1912, near the spot where he made aviation history, Cal Rodgers's plane collided with a seagull 100 yards west of Pine Avenue pier. Just before he crashed, Rodgers told Earl Daugherty he would "be back in a few minutes" to finish negotiating the sale of an airplane. Rodgers's death marked the 22nd American aviator to die while flying. In 1917, Frank Champion fell to his death while flying over Kochi, Japan. Champion's widow and 6-year-old son resided at 416 East Poplar Street in Compton.

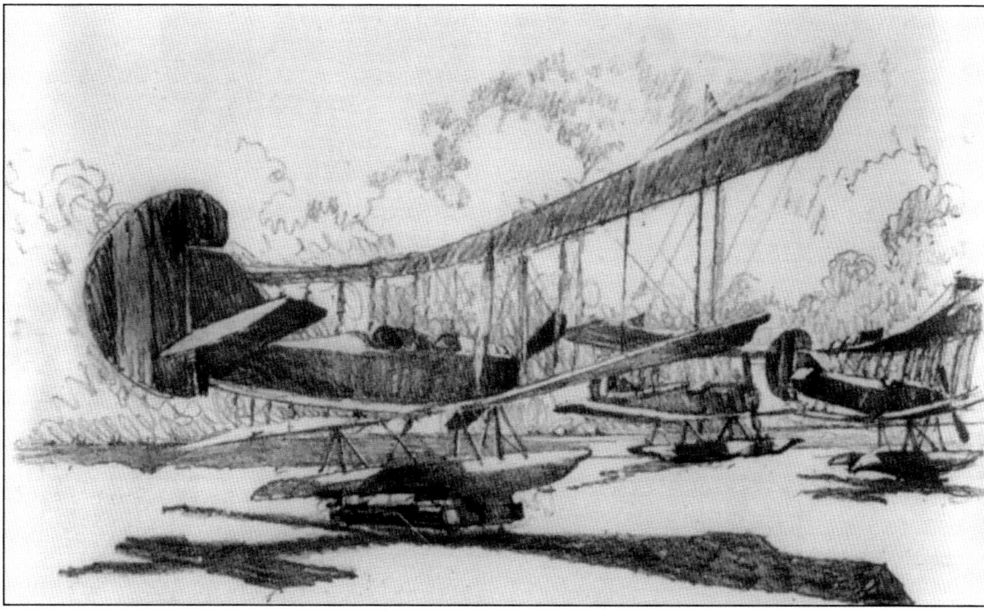

Hydroplanes from Long Beach. These photographs show early hydroplanes that eventually would carry passengers between Long Beach and Catalina. Glenn Martin employed Donald Douglas as an engineer and developed a hydroplane in Santa Ana. He challenged Long Beach aviator Frank Shaffer to fly to Catalina on May 10, 1912. Shaffer had been Cal Rodgers's mechanic and entered the race to help set a U.S. record over water and to raise money for a monument to Rodgers. Shaffer planned to take off near the bathhouse in Long Beach, fly to Avalon, and return with a sack of mail. Martin took off from Balboa Island and completed the 33-mile flight in record time, besting Louis Bleriot's 1909 crossing of the English Channel. Shaffer cancelled citing strong winds, choppy waves, and a smashed skid his plane sustained when he took it to Alamitos Bay. The final blow to Shaffer's flight was the withdrawal of his financial backers after Martin's success. Horace Kearney, a flyer at the Dominguez Air Meet, attempted to fly a hydroplane in December 1912 from the Long Beach port area to San Francisco. He crashed and died near Point Fermin. (Both, LOC.)

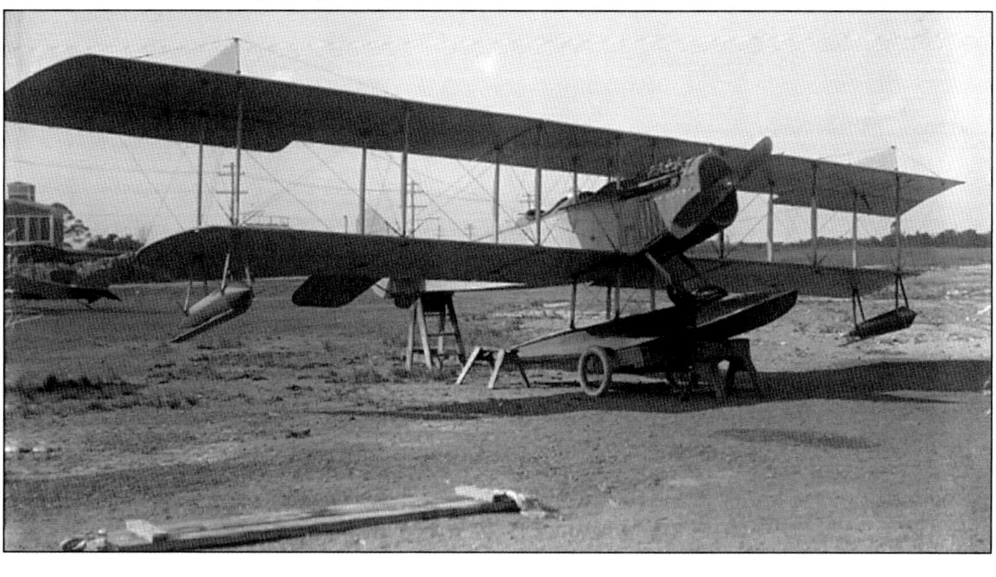

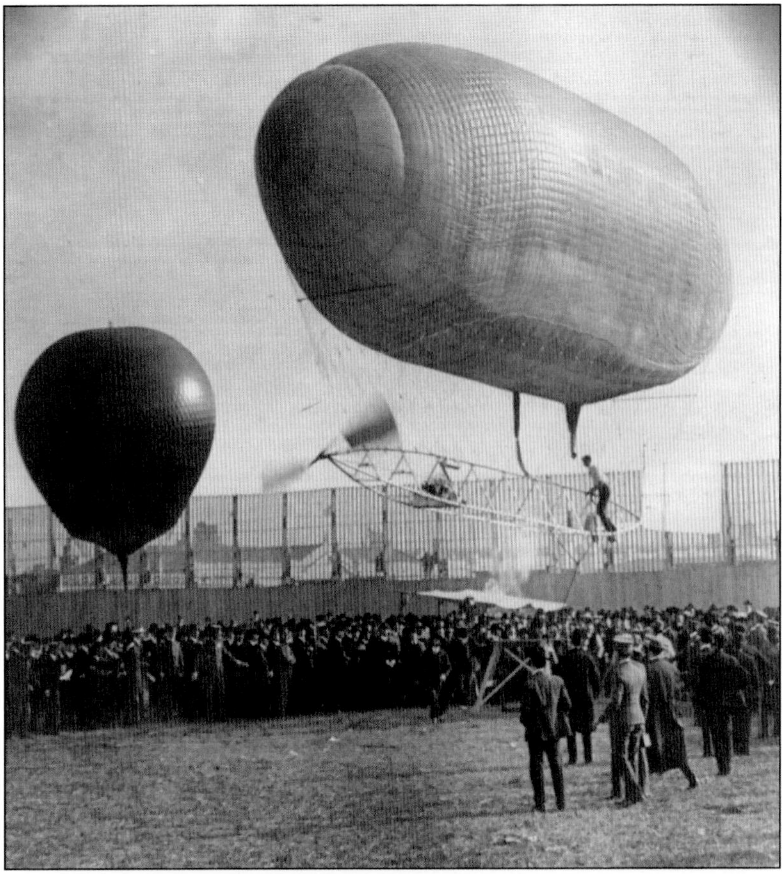

Airship Service. Roy Knabenshue (pictured in his airship, the *California Arrow*) raced his airship against Lincoln Beachey (below) at the 1910 Dominguez Air Meet and worked as a promotional manager for the Wright Company. Long Beach press reported airships (dirigibles) flying from Pasadena as early as 1913. Knabenshue opened his airship service in 1914, flying between Pasadena, Los Angeles, Santa Monica, and Long Beach. (Both, LOC.)

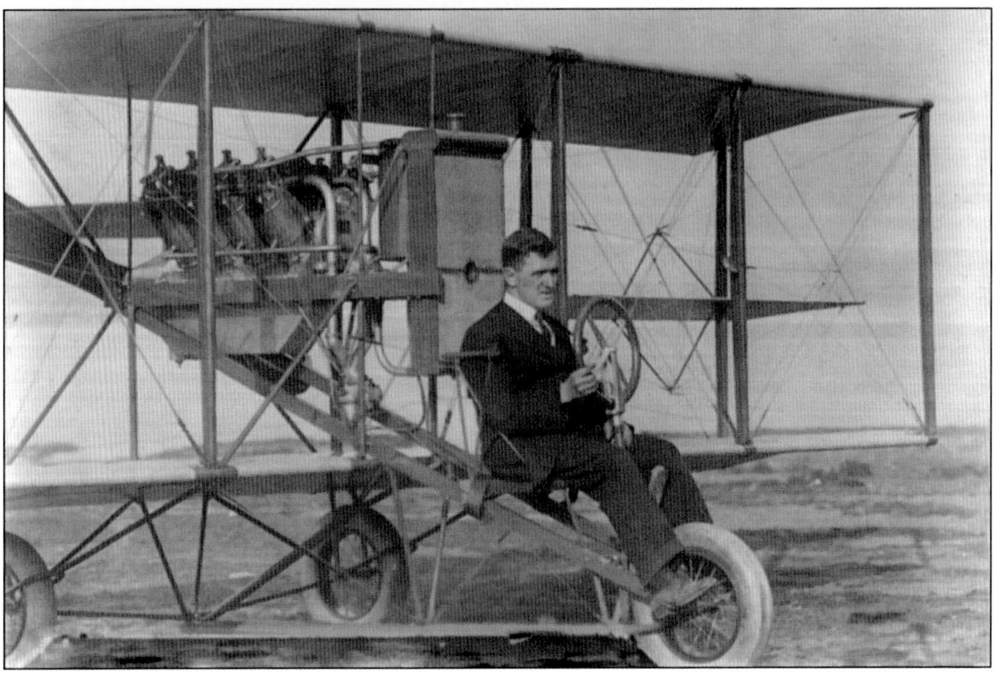

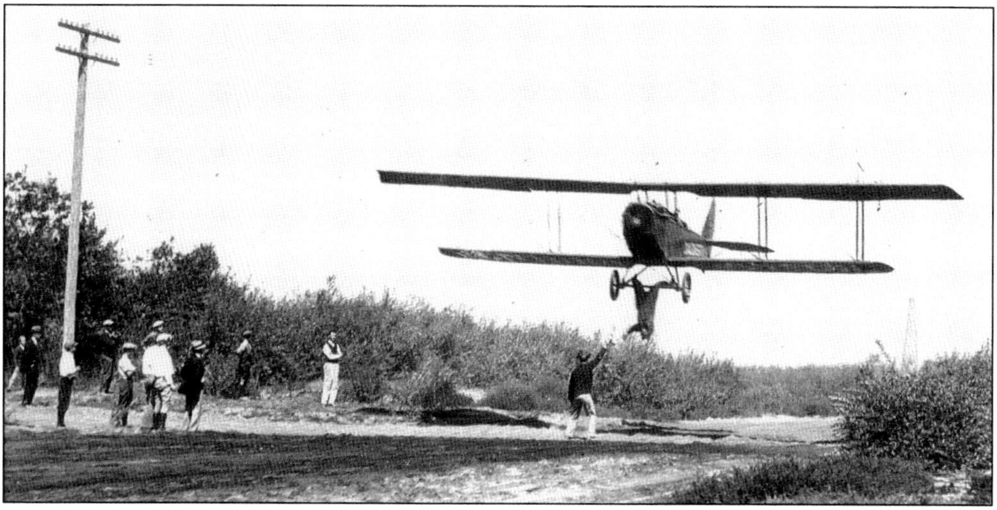

Silent Movies. Early aviators, including Earl Daugherty (piloting the plane pictured), performed their aerobatic stunts for the nascent movie industry. In the early 1900s, Long Beach was home to the Balboa Amusements Producing Company (also known as Balboa Studios), one of the largest movie studios in California. Located on 8 acres at Sixth Street and Alamitos Avenue in Long Beach (and 11 acres in Signal Hill), the studio featured the first aerial moving picture studio in the world. Stars like W. C. Fields (whose home still stands on Ocean Boulevard), Slim Pickens, Buster Keaton, Mabel Normand, and Fatty Arbuckle were a common sight in Long Beach. Several of Daugherty's flights were filmed by the Sunset Moving Picture Company of Los Angeles and used as advertising for travel to eastern cities. The Long Beach movie studio closed in 1923. (Above, KL; below, LOC.)

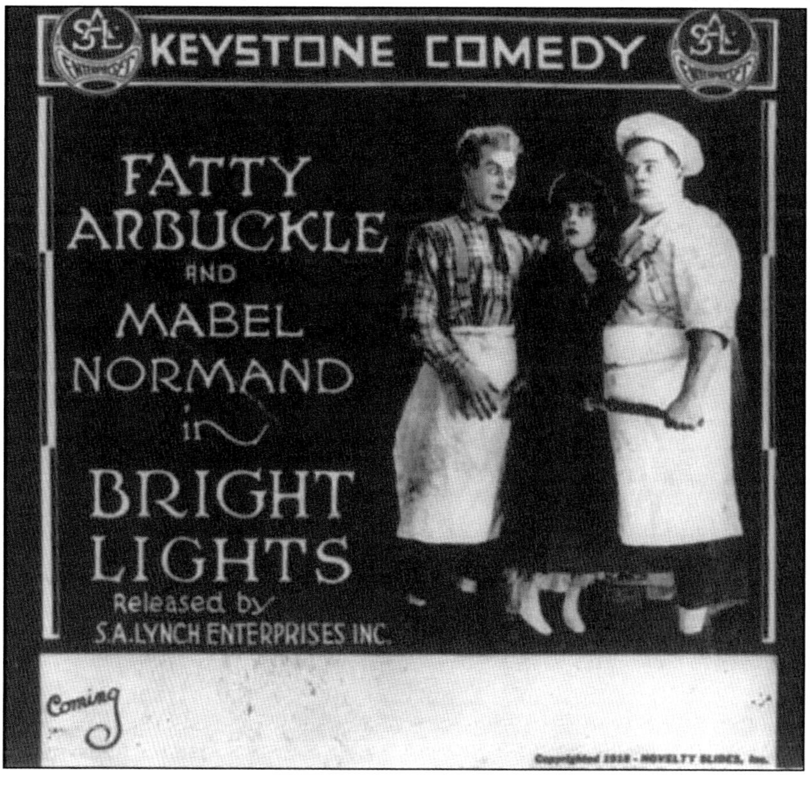

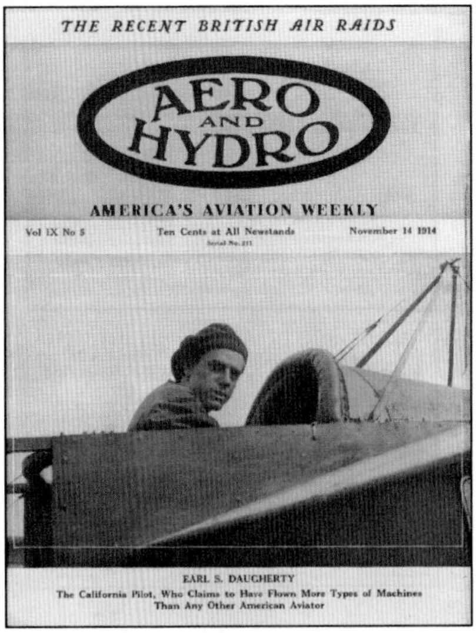

Every Type of Machine. Daugherty flew more types of air machines than any other American aviator, as was noted in the November 14, 1914, *Aero and Hydro* magazine. Among the planes flown were a Pontkowski-Lichorsik-Vought Tractor Biplane, which he crashed twice in Illinois, and a Hillery Beachey Experimental Biplane. Hillery was the brother of noted aviator Lincoln Beachey.

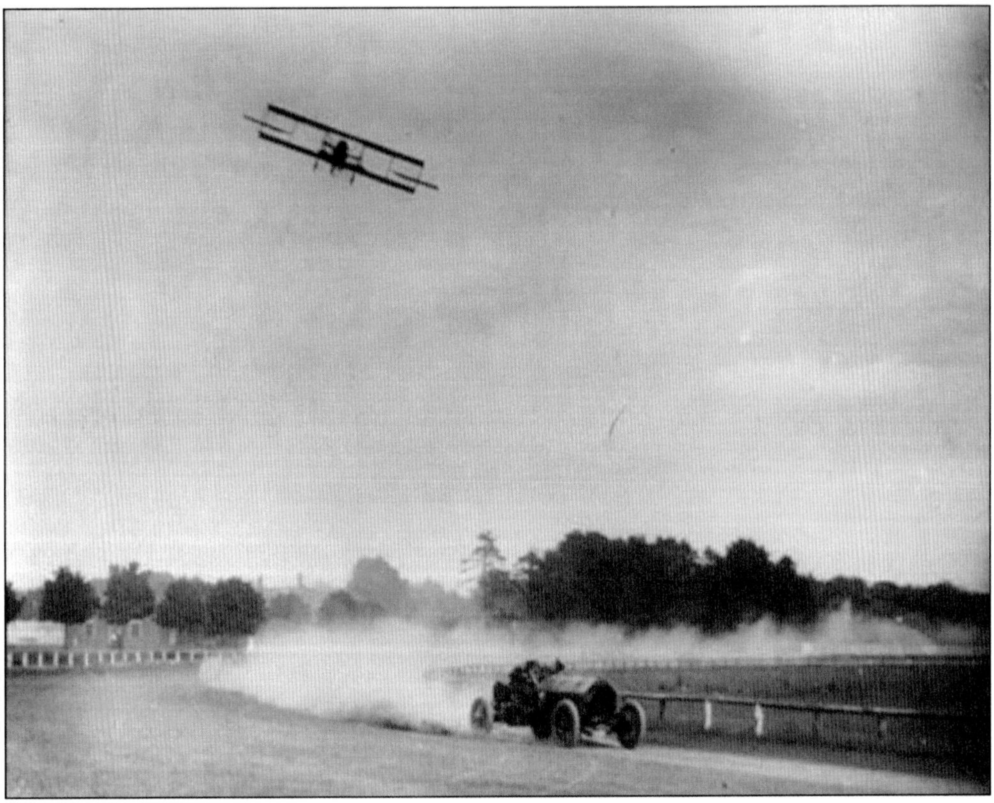

Racing Barney Oldfield. In 1914, Daugherty raced against his business partner and famous racecar driver, Barney Oldfield, who drove an automobile at the Hawthorne Race Track. Their race repeated the run (pictured) between famed aviator Lincoln Beachey and Oldfield in 1912. In 1915, a "soothsayer" at the Pike told Daugherty to "quit flying or die." Daugherty had recently fallen 50 feet near the Pacific Bathhouse, striking the beach so hard he cut his face and pushed a tooth into his jaw. After receiving the warning, he turned down an offer from Gen. Pancho Villa asking him to serve as an aerial scout in Mexico. (LOC.)

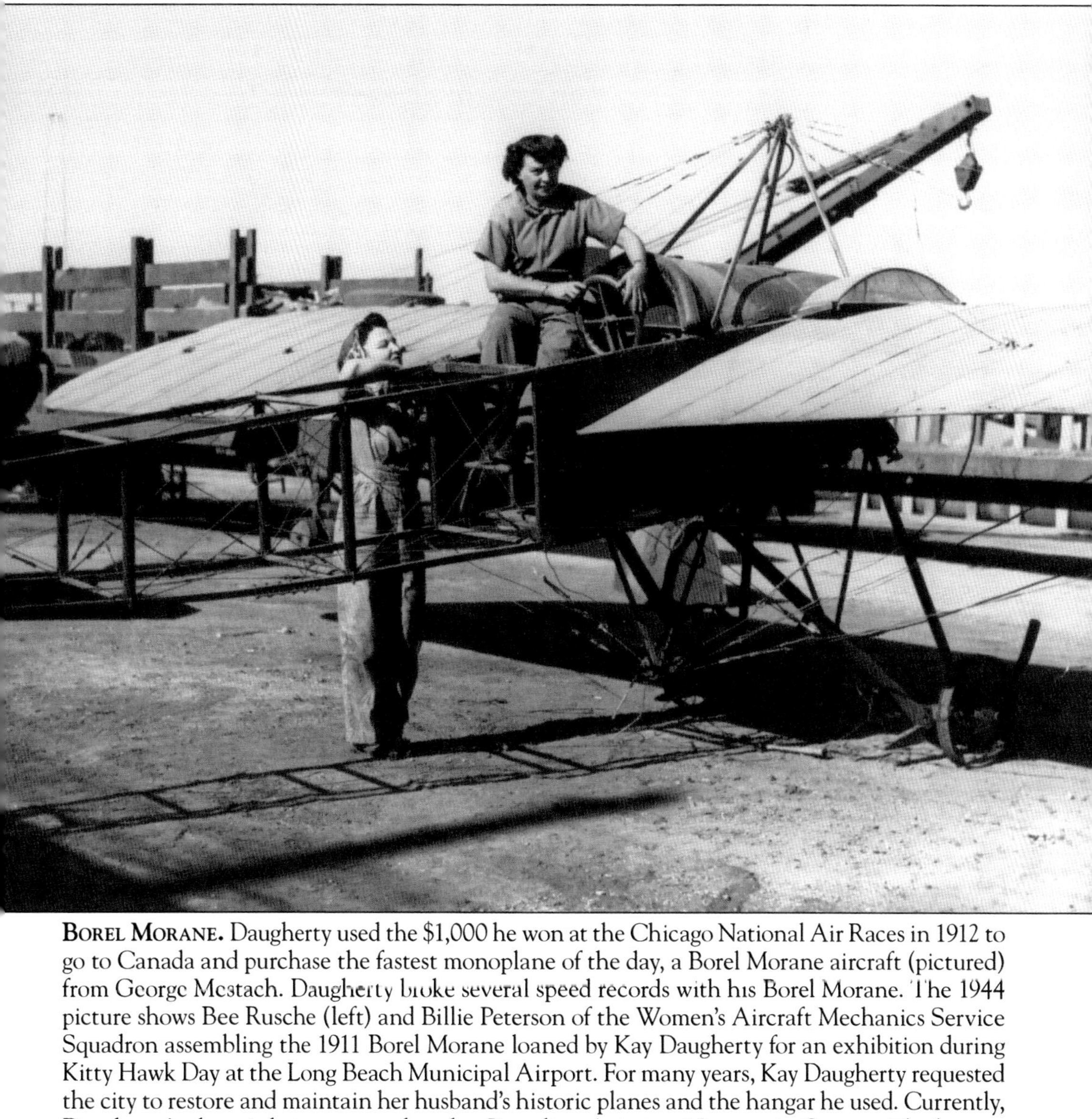

Borel Morane. Daugherty used the $1,000 he won at the Chicago National Air Races in 1912 to go to Canada and purchase the fastest monoplane of the day, a Borel Morane aircraft (pictured) from George Mestach. Daugherty broke several speed records with his Borel Morane. The 1944 picture shows Bee Rusche (left) and Billie Peterson of the Women's Aircraft Mechanics Service Squadron assembling the 1911 Borel Morane loaned by Kay Daugherty for an exhibition during Kitty Hawk Day at the Long Beach Municipal Airport. For many years, Kay Daugherty requested the city to restore and maintain her husband's historic planes and the hangar he used. Currently, Daugherty's plane is being restored at the Canadian Aviation Museum in Ottawa. The hangar was demolished in 1957.

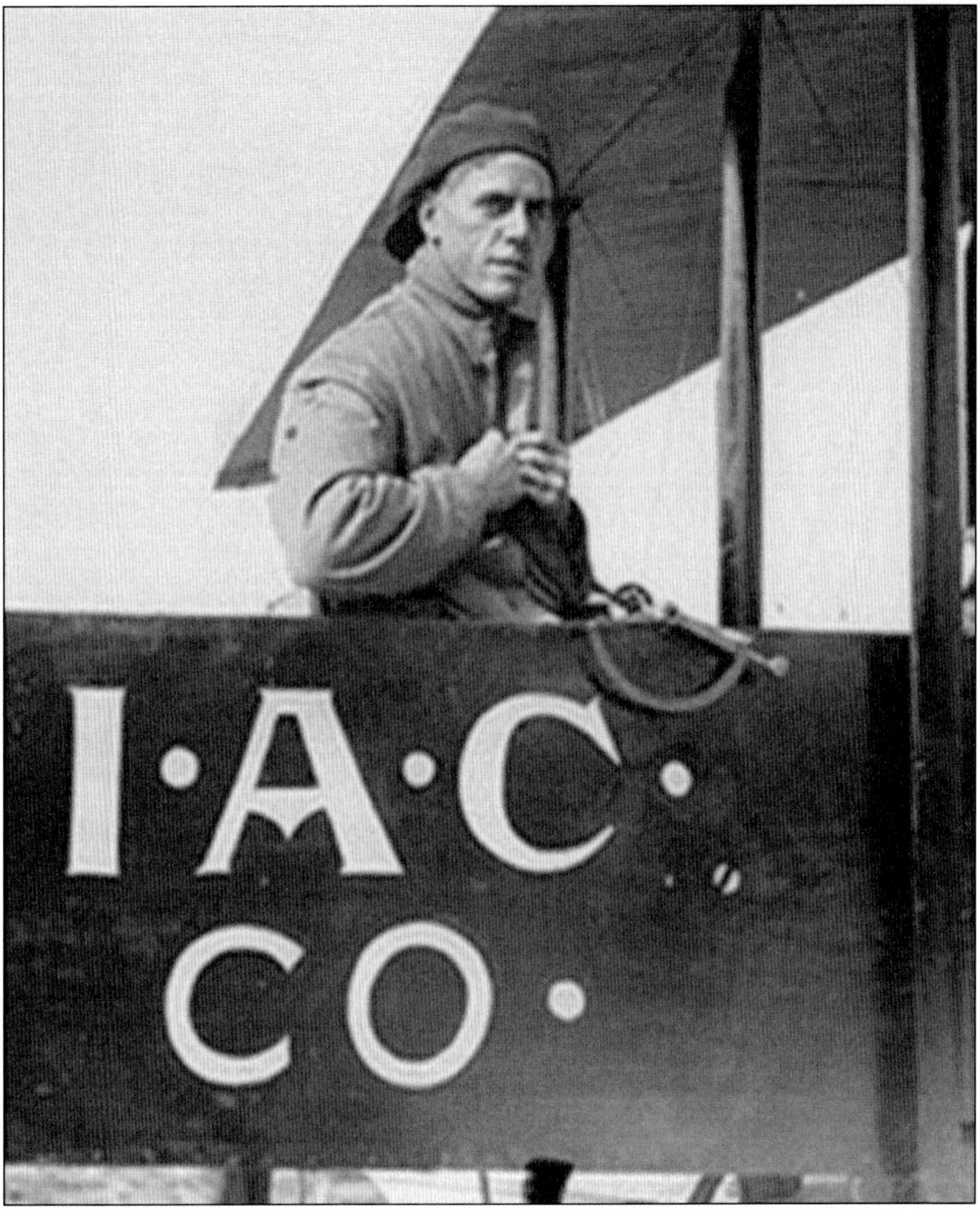

CHIEF TESTER. In between stunt flying in Long Beach, Daugherty went to work for the Frederickson Engine World Motor Company in Bloomington, Illinois. He also worked as the chief tester and first assistant designer at the Chicago Aero Works, which produced "air machines" for England, France, and Belgium. Illinois press noted Daugherty held several patents and his "brilliant flights have been exceeded by his genius in the manufacture of flyers of various kinds." On November 4, 1916, Long Beach announced the building of six airplane hangars on the beach to accommodate local aviators and the military. Residents circulated a petition to bar aviators from flying over Long Beach because cattle and other farm animals were running from the planes. In protest, Daugherty moved his flying exhibition to Seal Beach for two weeks.

Four

World War I Changes Aviation

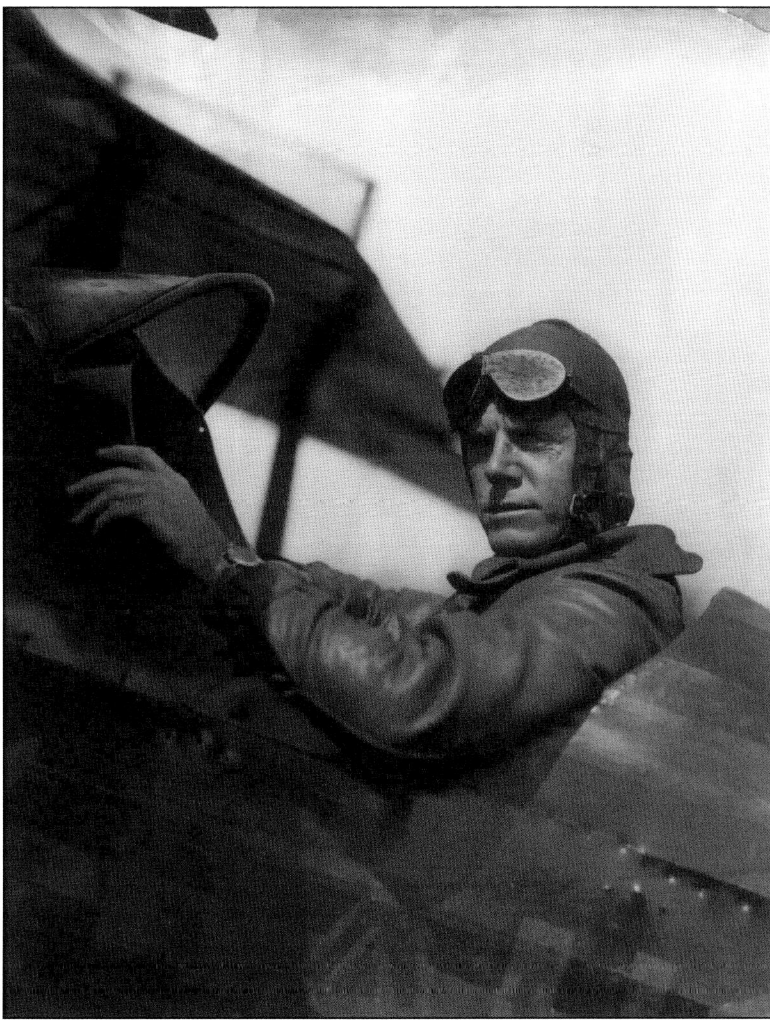

Ace Instructor. In 1916, there were only 200 licensed pilots in the United States. Earl Daugherty announced he was establishing a school of aviation and rebuilding his "tractor" plane in 1917. However, he accepted an offer to teach for $3,600 a year and was given a commission as second lieutenant, reserve military aviator in the Aviation Section, Signal Corps, U.S. Army, serving from July 7, 1917 to March 8, 1919.

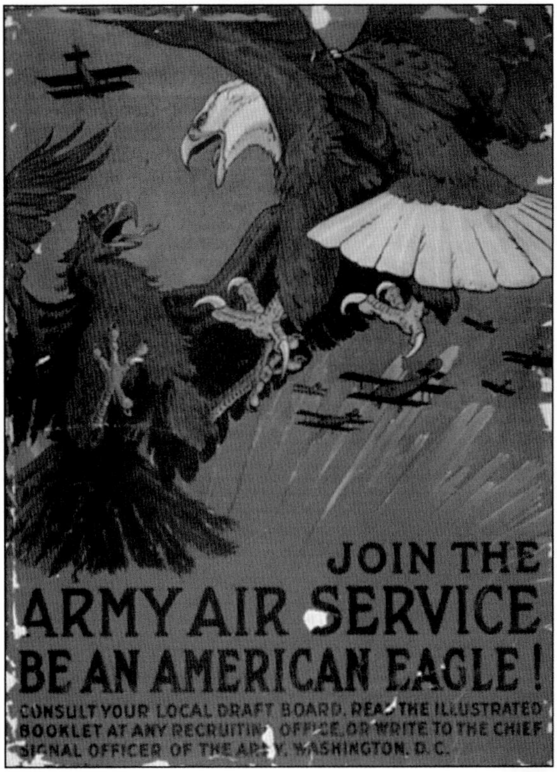

Army Air Corps. In 1914, the Aviation Section of the U.S. Army was formed. It was renamed the U.S. Army Air Service in 1918, the Army Air Corps in 1936, the Army Air Forces in 1941, and the U.S. Air Force in 1947. Although civilian flying was put on hold with the start of the United States' involvement in the war, military aviation took off. Thousands of men signed up to learn how to fly. The average lifespan of a World War I flyer was less than three weeks in combat. Those who survived returned to the states and made aviation their lives after the war. (Both, LOC.)

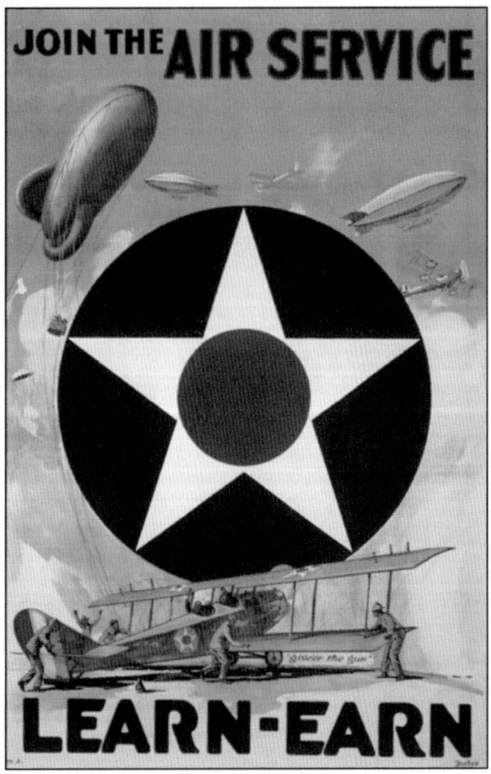

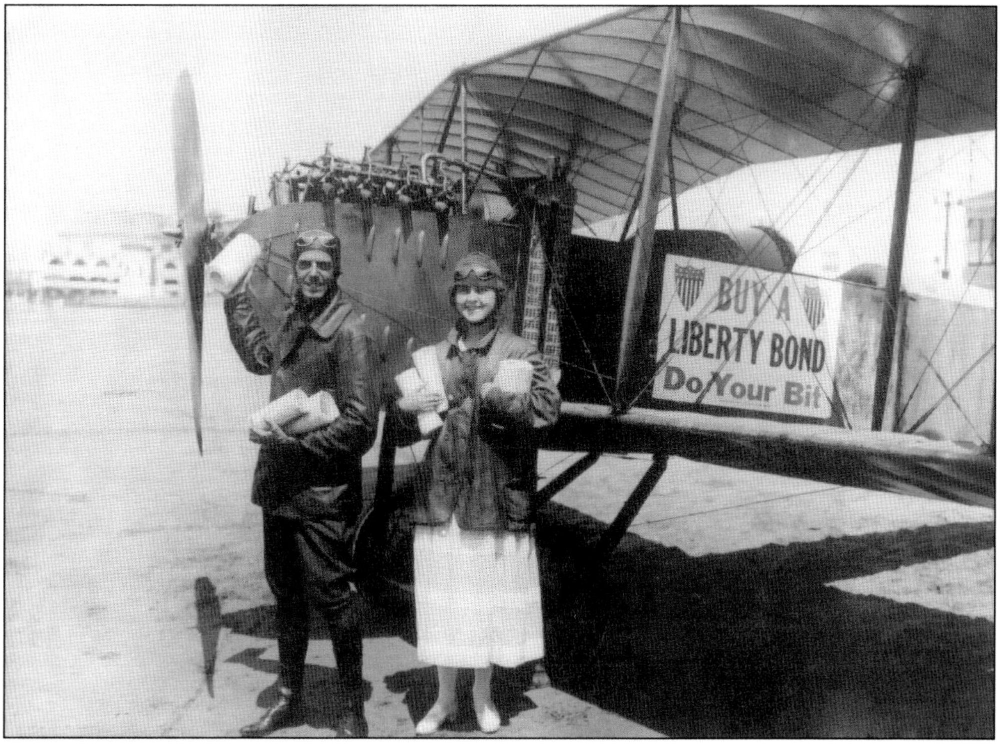

LIBERTY BONDS. Earl Daugherty and Lucille Crenshaw of El Paso, Texas, flew over Long Beach in June 1917, dropping flyers appealing to residents to buy Liberty bonds. During his military service, he continued performing air stunts in Long Beach. His flying was so popular that he was unable to land on the sand on June 1918 because of the crowds.

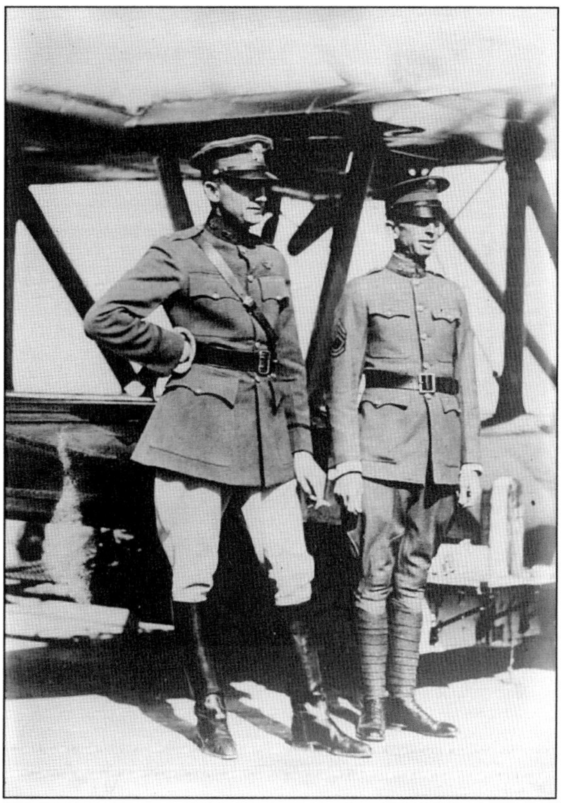

DAUGHERTY'S PUPILS. Daugherty served at both Rockwell Field, North Island, near San Diego and March Field in Riverside. He trained many important aviators, including army lieutenant R. L. Maughan, 1922 Pulitzer Trophy winner at the National Air Race at Selfridge Field; marine lieutenant D. W. Tomilson, member of first naval aerobatic team in 1927; and Lt. Lowell Smith (pictured left), one of the flyers of the Douglas World Cruiser around the world in 1924. (LOC.)

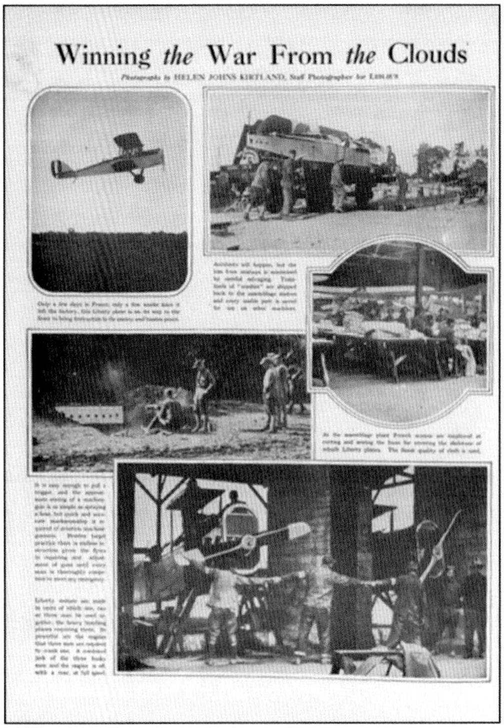

Winning *the* War From *the* Clouds

Airplane Uses. World War I proved airplanes could be used as a weapon and as an efficient, dependable, and swift means of communication and transport. Daugherty recognized the potential of aviation as a commercial venture for providing short flights, giving exhibitions, carrying passengers and freight, and taking aerial photographs for mapping and surveying. Flyers knew "more landing fields and better landing fields" was "the biggest problem in aviation," as noted in the January 1920 issue of *U.S. Air Service* magazine. The army trained pilots on the beach in Long Beach before Daugherty established the city's first inland field. (Above, LOC.)

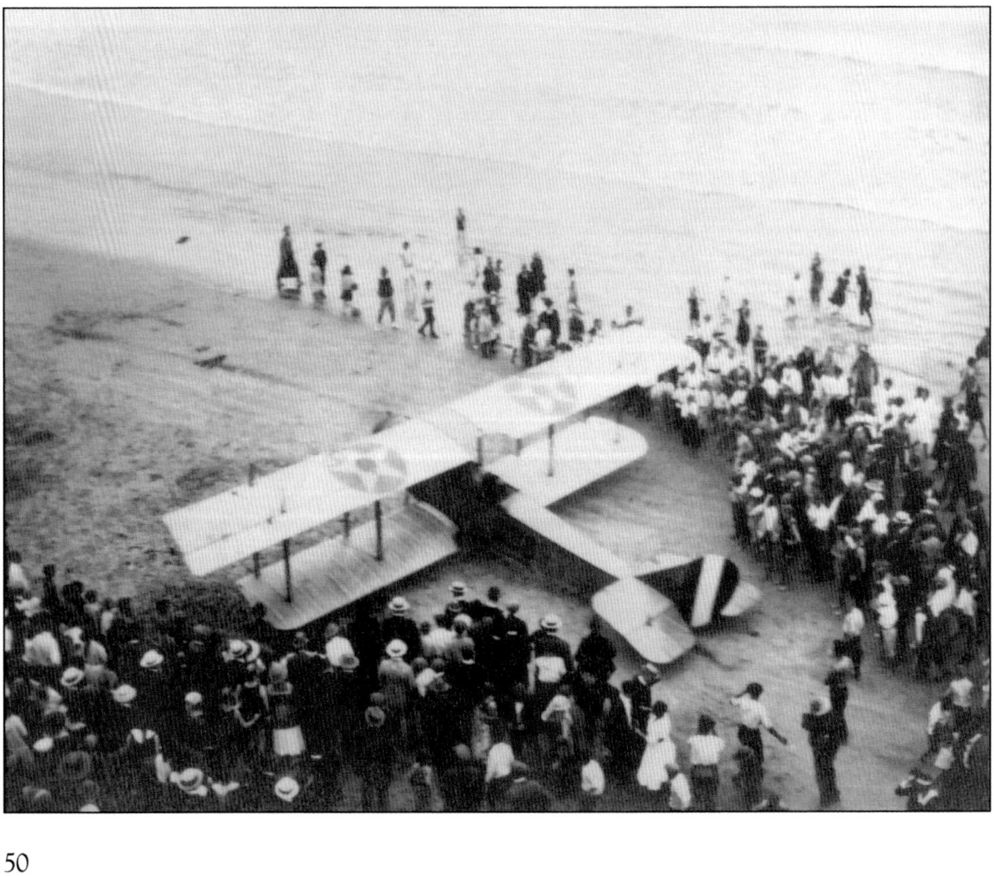

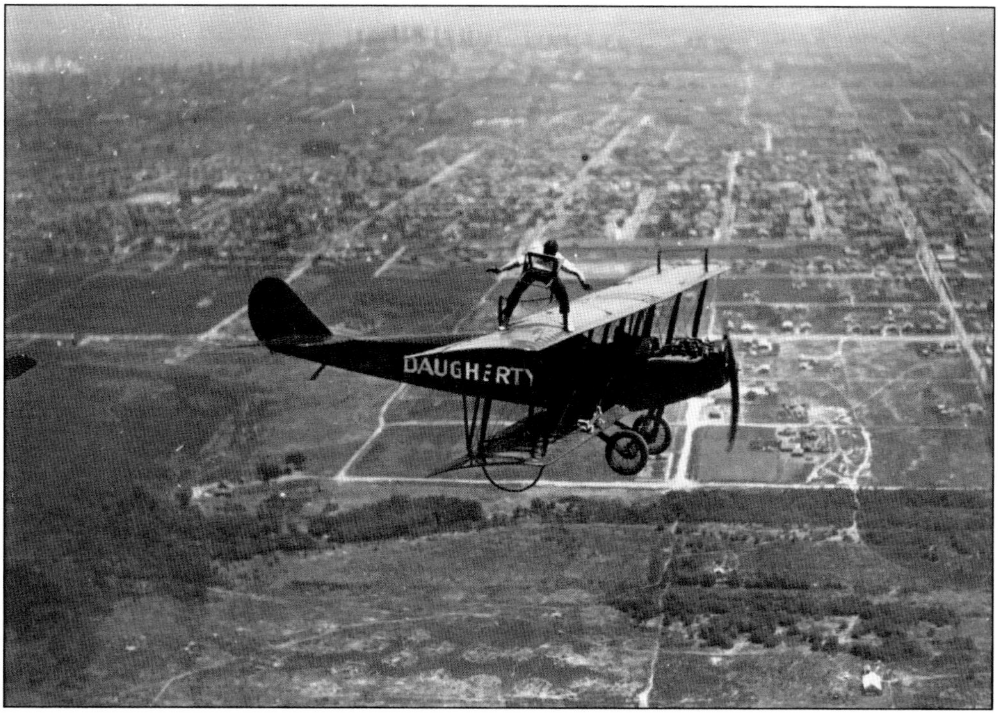

Best-Known Stunt Pilot. Daugherty returned to Long Beach as the oldest active flyer and best-known stunt pilot in the United States, setting the record of 72 continuous loops in the air. He also performed the highly dangerous stunt of maneuvering his plane around his friend (and member of the Long Beach Chamber of Commerce) Wayne Abbott, while Abbott parachuted from the plane. Daugherty and silent film actress Bessie Love also set the record altitude of 11,500 feet. Daugherty performed the stunt with Love to establish a new record flying with a "lady passenger," while being filmed by Gaumont Weekly Motion pictures. Plane design included stronger engines and sturdier construction, making it possible for higher altitudes and faster speeds. Stunt flying was an inexpensive, thrilling means of entertainment, glamorized in movies.

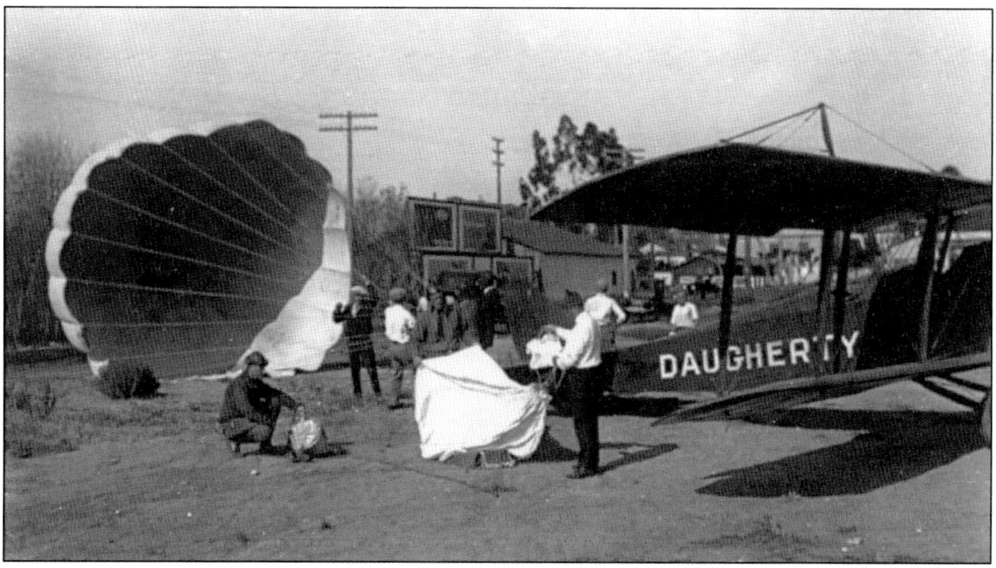

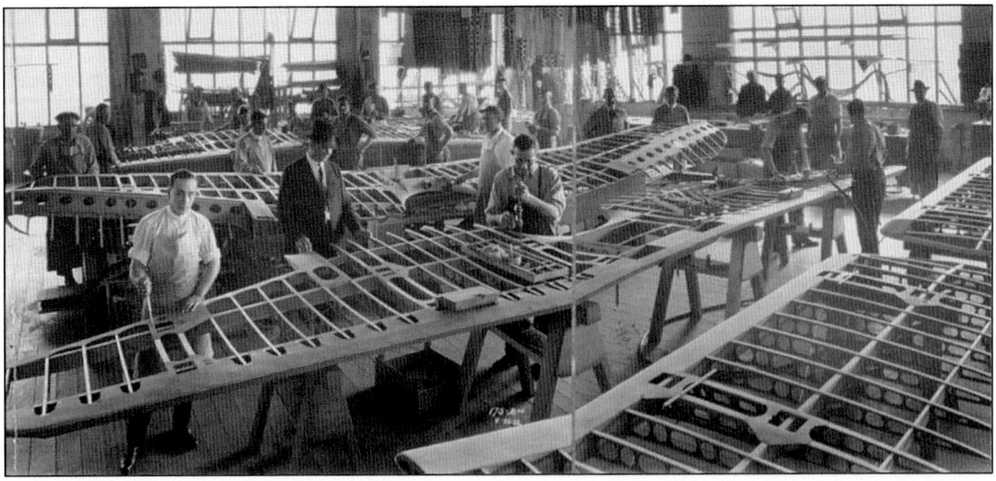

Golden Age of Aviation. The years 1918 through 1938 are called the "Golden Age of Aviation." When World War I ended, a surplus of military planes and trained pilots were available for civilian use. The Roaring Twenties and the entertainment provided by stunt "barnstormers" traveling the United States and appearing in movies added to the excitement. Plane design changed from wood-and-cloth contraptions (above) to enclosed metal planes powered by strong engines. Commercial investors offered lucrative cash prizes for speed contests, and flyers made aviation firsts. Airlines carried passengers, and the military expanded its demands for airplane manufacturing. Cities considered the airport a municipal utility. Aviation even influenced fashion, as shown in the photograph below of an "airship hat." (Both, LOC.)

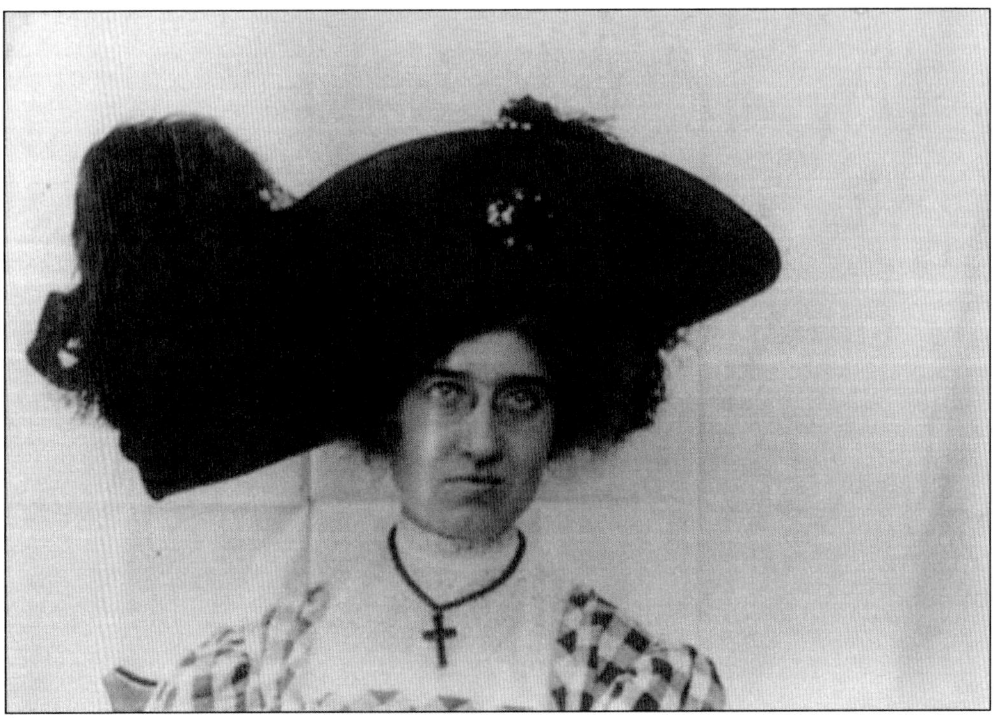

Five

Long Beach Gets an Airfield

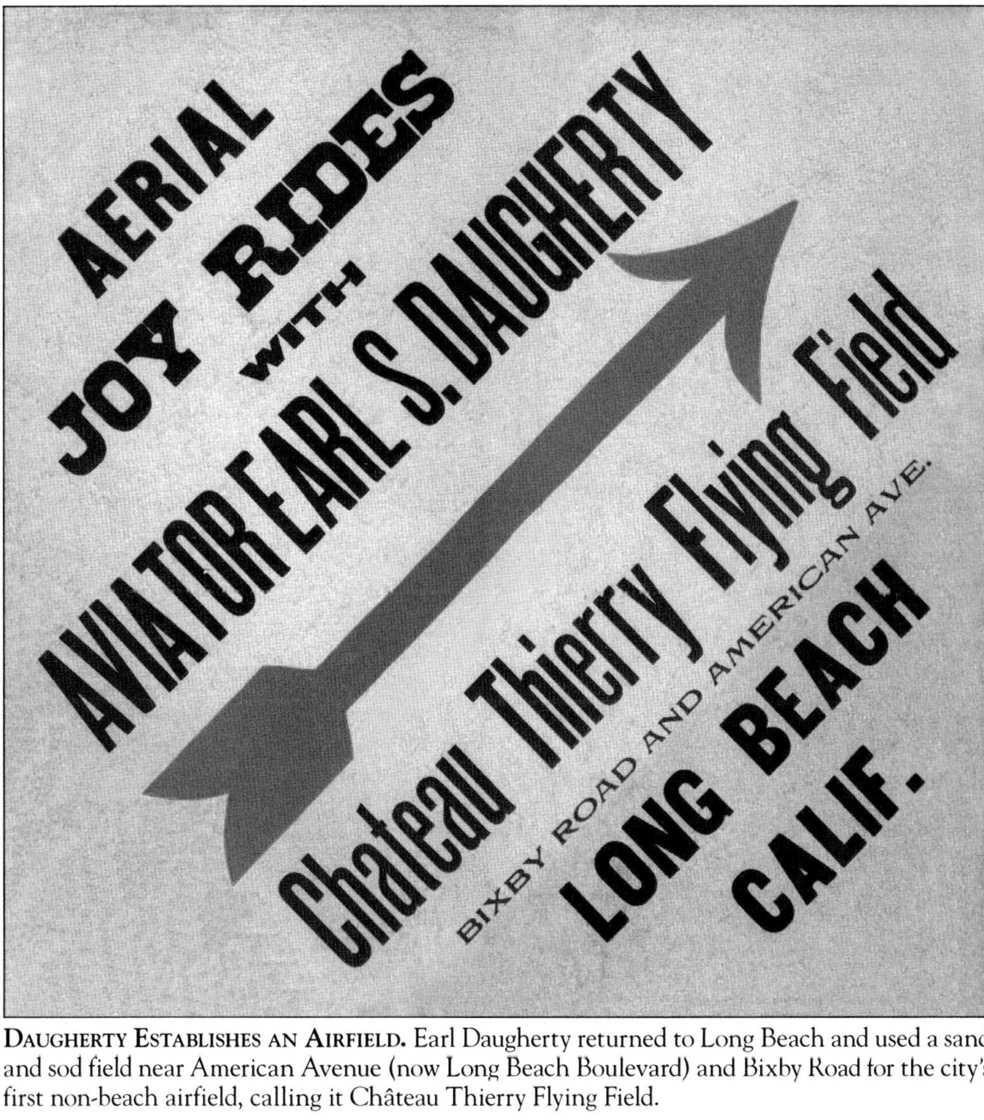

Daugherty Establishes an Airfield. Earl Daugherty returned to Long Beach and used a sand and sod field near American Avenue (now Long Beach Boulevard) and Bixby Road for the city's first non-beach airfield, calling it Château Thierry Flying Field.

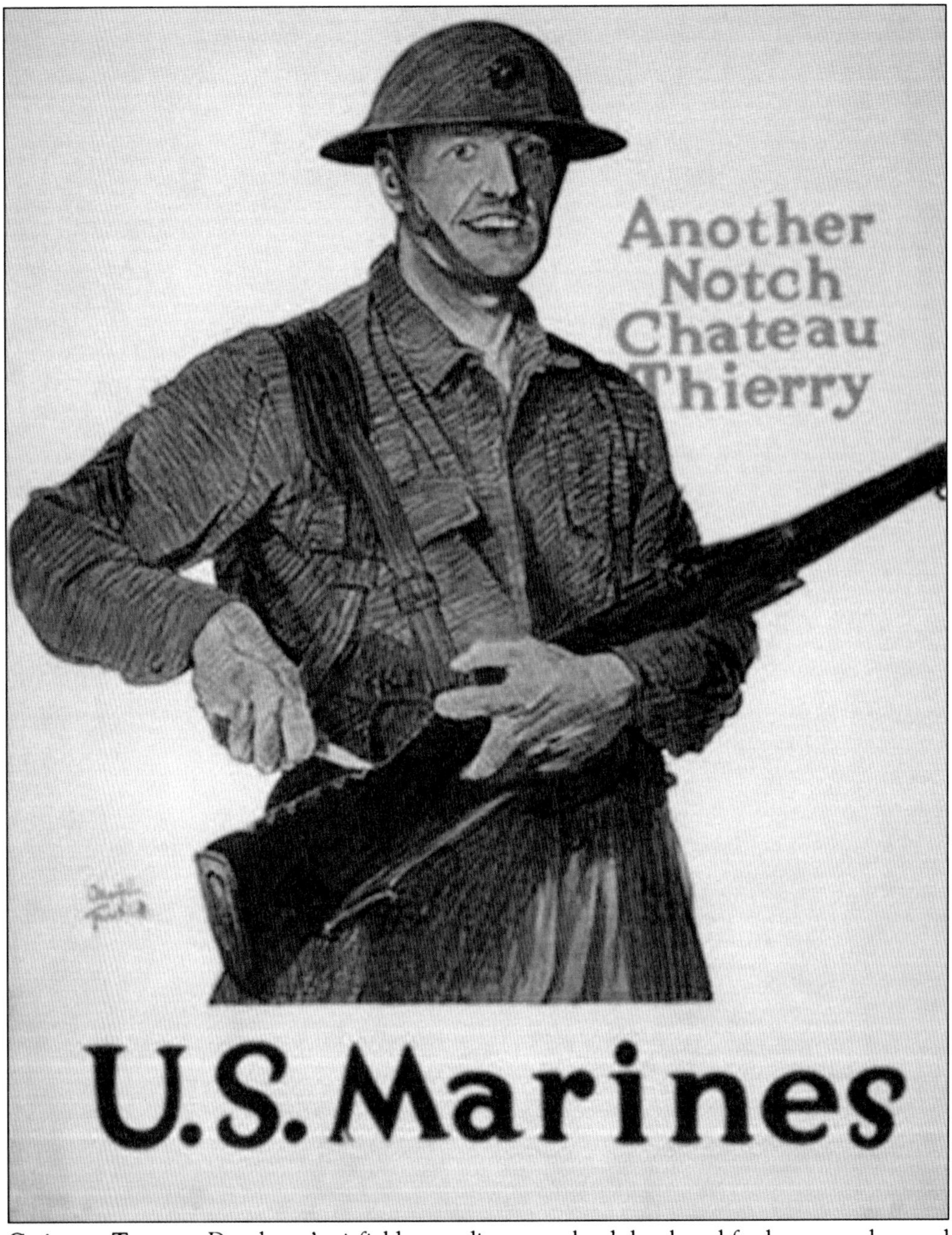

Château Thierry. Daugherty's airfield was adjacent to land developed for homes and named Château Thierry, after a beautiful area of France. Château Thierry was also the site of the first military battle in which American troops, under the direction of Col. Billy Mitchell, were supported by airplanes. Daugherty's student pilots were credited with downing 15 "Hun" planes. The battle was bloody, but troops were able to push the Germans back because of the planes. People came from miles away to watch Daugherty and the "wing-walkers" who worked with him perform their dangerous and daring stunts. Daugherty's "air rodeo" and "air circus" featured Jenny races, relay races, upside-down contests, and fireworks. Daugherty hoped to bring to Long Beach the same type of excitement that was storming across the nation. Daugherty charged $5 for a 5-minute ride.

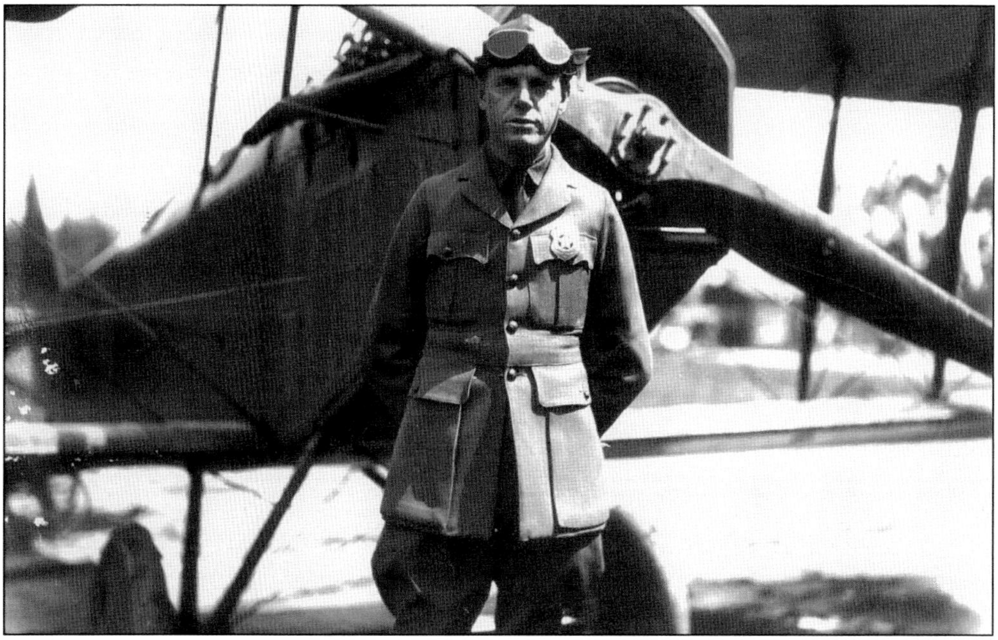

Police Pilot. From his Nieuport Speed Scout, Daugherty also engaged in search and rescue and other "police work," earning him the title "Knight of Aviation" and a badge as "Number 1 Police Pilot." In July 1919, he rescued Bobby Holman, three-year-old son of his friend and silent movie actress Nell (Shall) Holman.

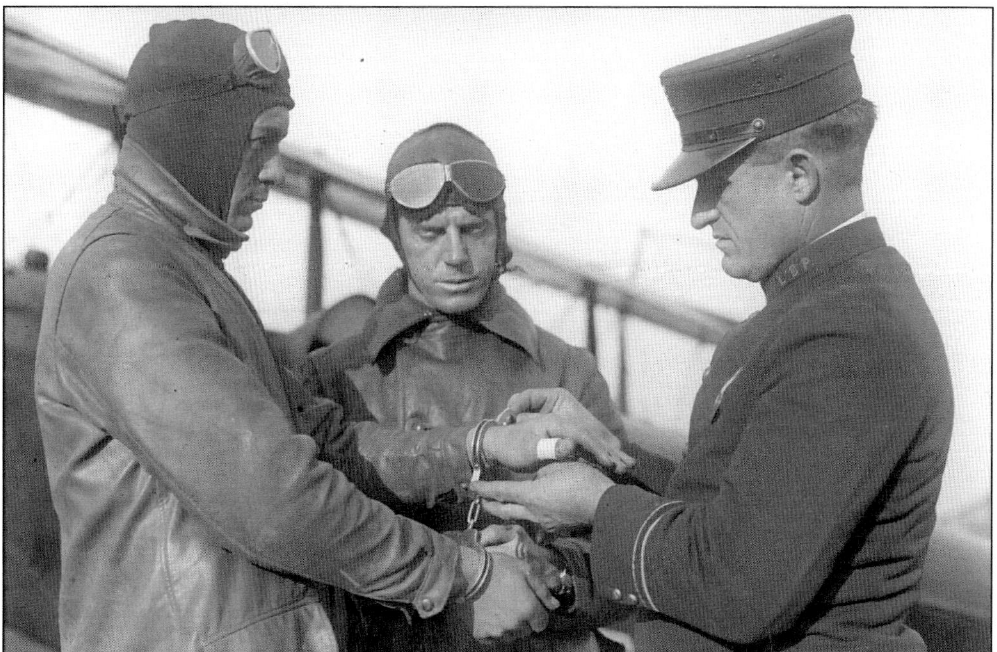

Stunts with Police Help. Daugherty assisted the Long Beach Police by dropping "air messages" concerning what he witnessed from the air. The police in turn helped him by hand-cuffing Wesley May, his partner for a stunt involving May walking across the wings of Daugherty's plane while in the air. (AC.)

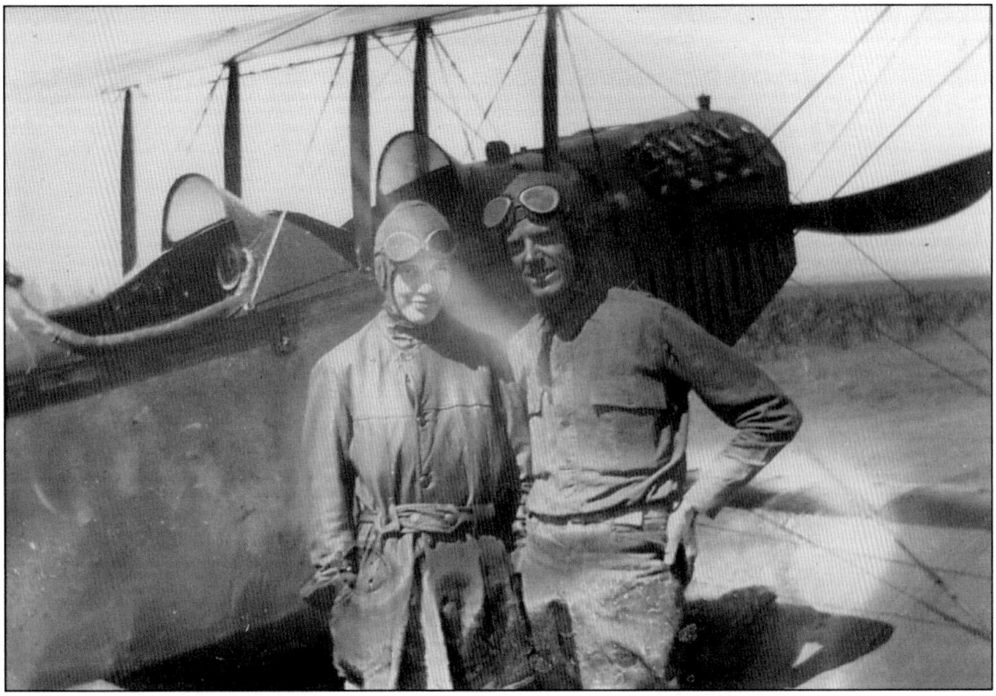

EARL S. DAUGHERTY, - - - AVIATOR

Aero Club of America, Licensed Pilot No. 87

CONTRACT.

THIS AGREEMENT, made and entered into this..........day of..........191.. by and between..........(..........) of the City of..........County, ... State of ....party of the first part and .... of the City of ....County of..........State of .... party of the second part.

WITNESSETH: That for and in consideration of the sum of one dollar by each of the parties to the other in hand paid, and for the other good and valuable consideration, both parties hereby agree as follows:

FIRST: The party of the first part hereby engages for flying purposes an experienced aviator, from the party of the second part for flying purposes at..........(..........) aeroplane for.......... ....days, beginning on ....191.. and ending on ....191.. and hereby agrees to pay the party of the second part the sum of $..........Dollars, in currency in the following manner, that is $..........cash immediately after the first day's flight, $..........cash immediately after the second day's flight, $.......... cash immediately after the third day's flight.

SECOND: The party of the first part hereby agrees to furnish a protected space and erect upon same a tent or hangar ....by ....feet in which the aeroplane may be housed, and furnish watchman to guard same at night. Agrees also to furnish local license for the flights and exhibition of said aeroplane, if required.

THIRD: For the purpose of permitting the aviator to make ascents from the ground with safety, said party of the first part agrees to furnish the entire infield of the race track and provide a clear and unobstructed flat space, appropriate for aviation, of not less than..........feet wide by..........feet long. This space is to be clear of all over-head wires, trees and other obstructions at one end, smooth and rolled, if in the opinion of the aviator it is necessary, properly policed so as to keep the people off the flying grounds during exhibitions, and to protect the property of the second part.

FOURTH: The party of the second part guarantees twenty minutes actual flying daily and agrees to make flights whenever the velocity of the wind is fifteen miles an hour or less. The length and duration of the flights to be at the discretion of the aviator, a successful flight to be from one to thirty miles in length and arising to a height of from fifty to five thousand feet.

FIFTH: The party of the second part agrees to make..........flights daily (except otherwise provided in this contract) or..........flights during the..........days of the meeting.

SIXTH: Should the weather be so inclement during the entire dates of the meeting as to prevent the party of the second part from making any flight or flights, then the party of the first part shall pay to the party of the second part the sum of $..........

SEVENTH: If through inclement weather the meeting should be postponed for one day, then the party of the second part agrees to make a flight, for which they will not receive any extra remuneration therefor.

EIGHTH: If any flight or flights are not given, owing to the fault of the aeroplane, or the aviator, on any specified day, then the aviator shall have the privilege of making two flights on the next favorable day, said extra flight to be considered as made on original date and paid for in accordance with the terms of this contract.

NINTH: Said party of the second part shall have the privilege of exhibiting the aeroplane within the hangar both before and after the flights, during the term of this contract, to which an exhibition fee may be charged and retained by the party of the second part, and the souvenir postcard privilege shall also be retained by said second party.

TENTH: It is understood and agreed between the parties hereto that should the aeroplane be broken into such shape as to prevent further flights during the life of this contract, the party of the first part agrees to waive any and all claims for damages that might arrive therefrom, and should the aeroplane be damaged beyond repair while making a flight, such flight is to be considered as a successful flight and is to be paid for according to the terms of this contract.

ELEVENTH: IT IS FURTHER AGREED between the parties hereto that if on account of accidents to, or illness of the aviator, damage to the aeroplane, or other unavoidable or providential circumstances beyond the control of the said party of the second part, said air flights shall not be made as hereinbefore written, then the said party of the second part shall not be held liable for loss or damage therefrom, and said second party shall reserve the right to replace with an experienced aviator to fulfill the unexpired part of this contract.

TWELFTH: IT IS FURTHER AGREED AND UNDERSTOOD that said first party assumes all liabilities to the general public, resulting from accidents occurring on the infield of the race track, or aviation field, and shall protect the second party therefrom.

WITNESS, the hands of the parties hereto, this..........(..........) day of ....A. D. 191..

....

By..........

....

By..........

Experience Is Insurance. Daugherty used this motto to attract passengers, especially prominent Long Beach residents, who flew with him on short flights over the city and surf. The 1919 photograph shows Mary Agnes Simpson, daughter of A. W. Simpson, who owned the Long Beach Sanitarium at Fifteenth Street and Linden Avenue, with Daugherty. He used a contract when performing exhibition flying and guaranteed at least "twenty minutes actual flying daily . . . whenever the velocity of the wind is fifteen miles an hour or less."

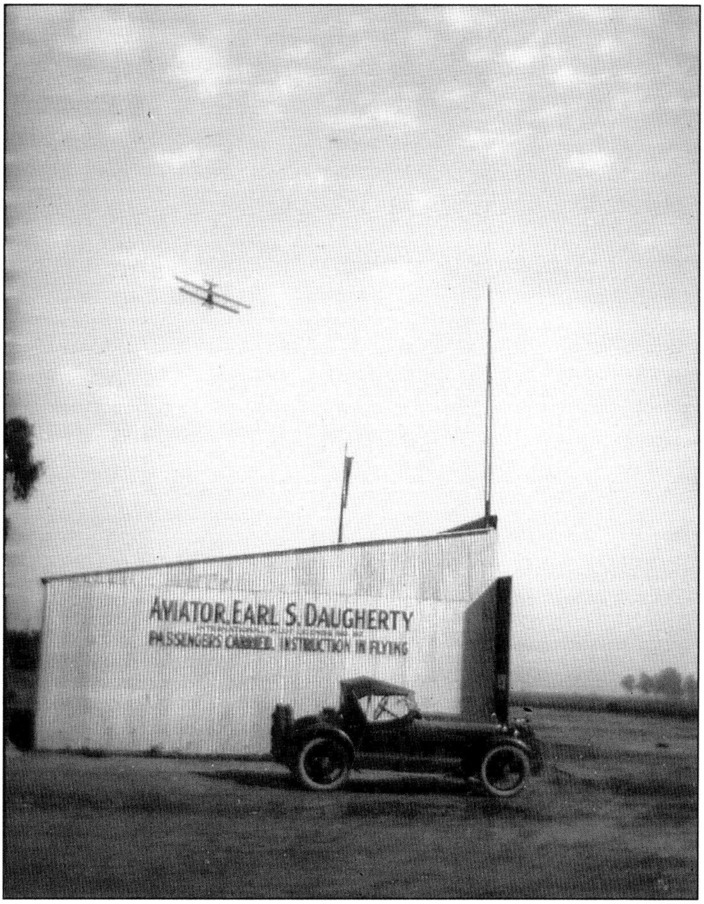

Growing Pains. By late 1920, Earl Daugherty ran a successful school of aviation and airplane operations at Château Thierry Field. Long Beach, its school system, and Daugherty needed land to expand. He moved to 23 acres at Long Beach Boulevard and Willow Street owned by his parents. Daugherty envisioned the new site as a large air terminus for both commercial and military aviation in Long Beach, a vision supported by the city council, which later used oil money to make needed improvements to create a municipal airport in an adjacent area.

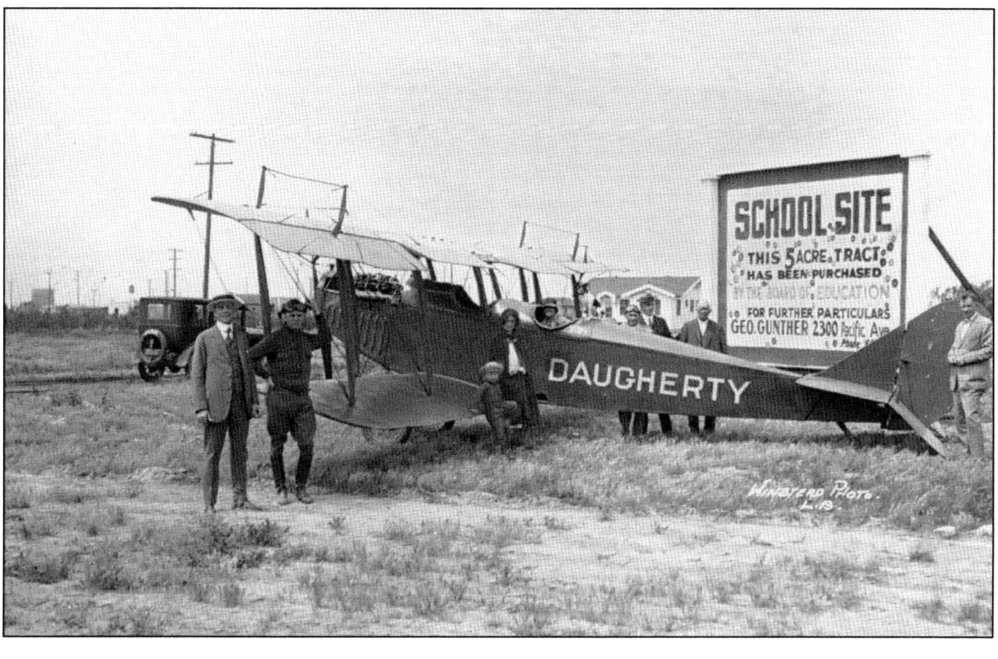

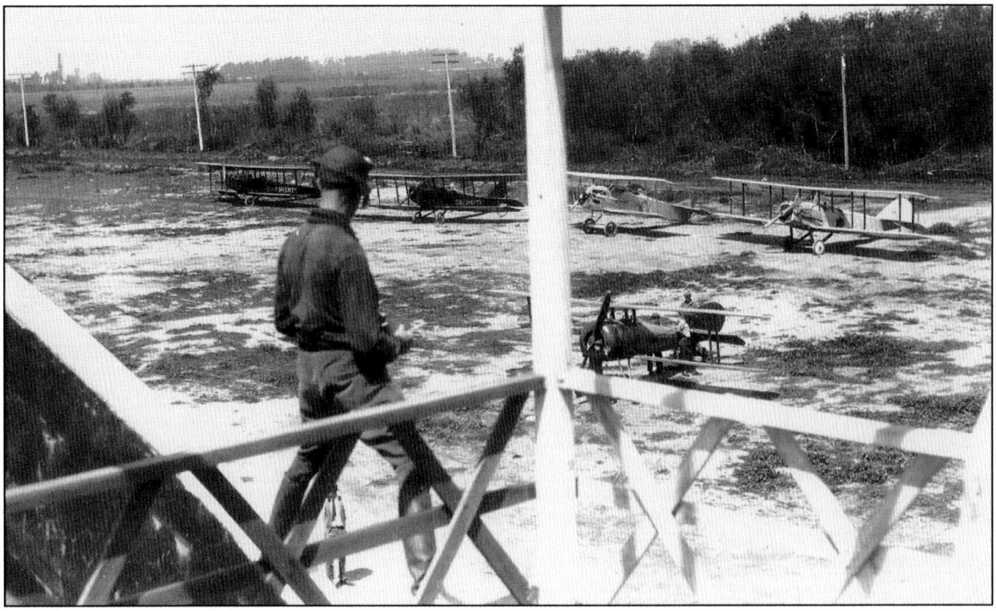

Air Circus. Earl Daugherty is pictured surveying the variety of planes on his airfield. Daugherty Field was one of the most famous of the 146 air terminals in the United States at that time because of its L-shaped runways measuring 1,400 feet by 400 feet and 1,350 feet by 270 feet. The field included two hangars with a capacity of holding "11 ships." Daugherty donated part of his aviation field to the Long Beach Elks for a baseball field. (Above, KL.)

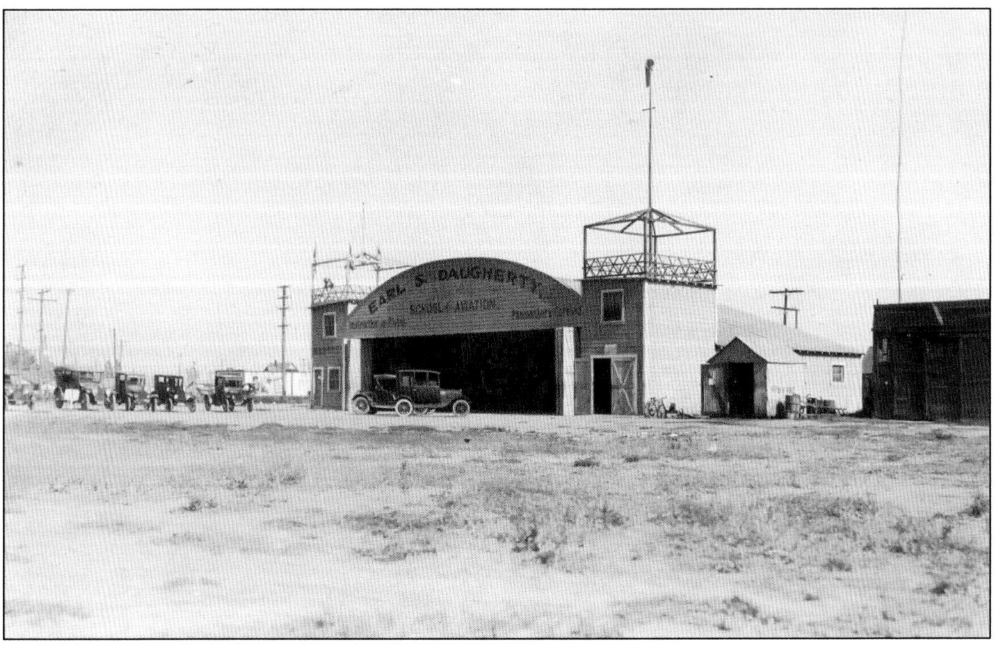

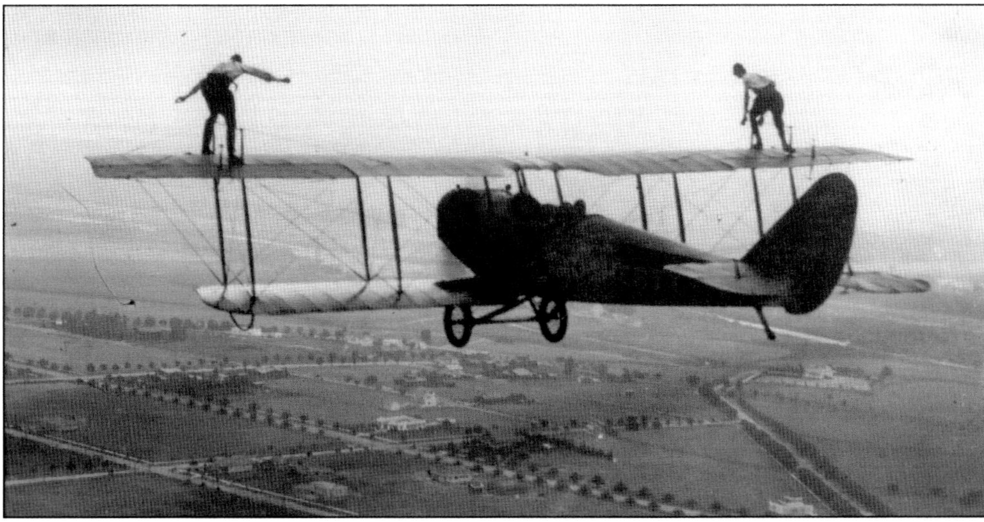

By the Skin of His Teeth. On September 5, 1920, Wesley May (left on wing) and Clarence "Ace" Bragunier (right) thrilled crowds by walking together on the wings of Daugherty's airplane. May then hung below with his teeth. Daugherty narrowly averted death the same year when his plane was carried out to sea by strong storm winds. Daugherty and the Aero Club of Long Beach sponsored a "National Winter Air Tournament" in December 1920. A 3,000-by-500-foot runway was constructed near the Willowville Trolley stop. The air meet included 25-mile races at speeds reaching 140 miles per hour. Daugherty, Jay Boyd, and Wesley May performed aerial stunts. More than 75,000 people attended, including a young Amelia Earhart. The *Los Angeles Times* remarked the exhibit's "show the largest of its kind held during the present year, surpassing both the New York and Chicago exhibits."

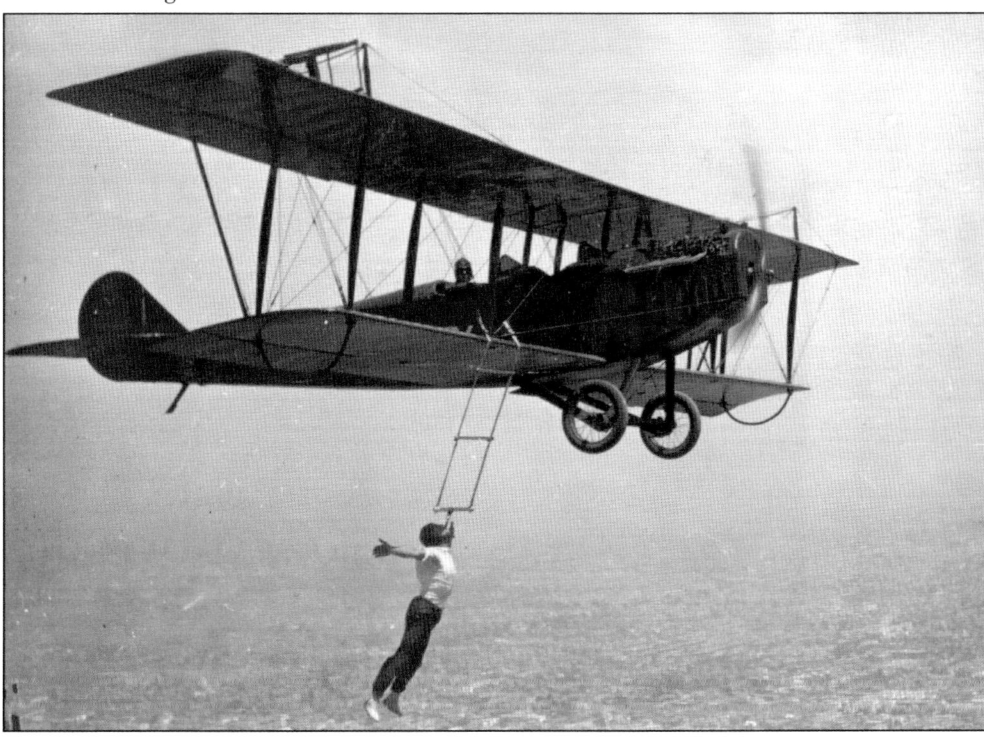

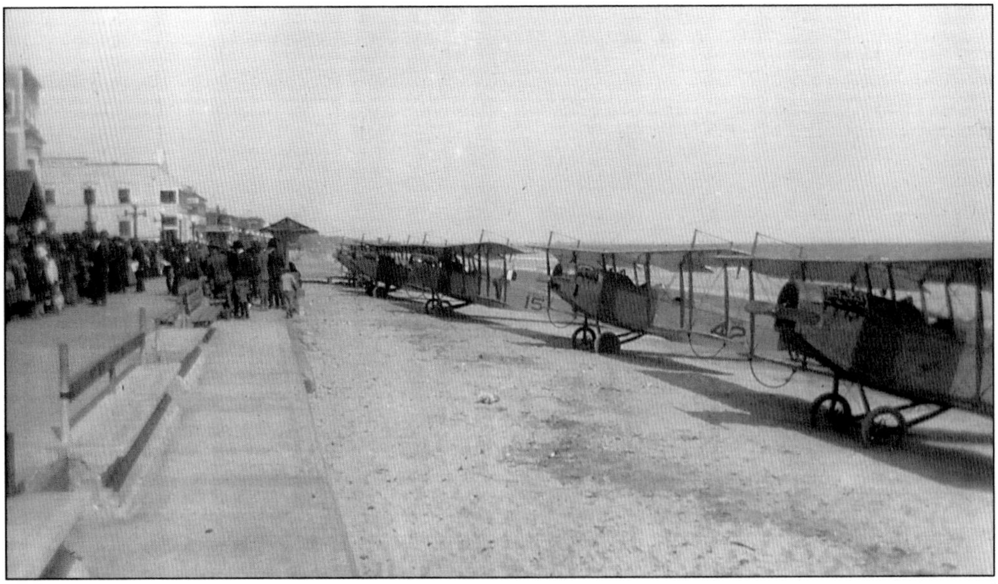

Surplus Military Airplanes. Surplus World War II Jenny (the Curtiss JN-4, shown on the sand in Long Beach) airplanes that sold for $100 and the lack of strict regulations for licensing flight instructors opened aviation to anyone. Daugherty capitalized on this trend by selling a plane with a package of flying lessons. He and Erving Diess (formerly with Syd Chaplin Aircraft) acquired rights to sell JN-4D, Oriole, and Seagull airplanes throughout Southern California and Arizona, which he advertised on the side of his hangar as the California-Curtis Airplane Company. In the 1920s, two-thirds of the 1,200 airplanes in business use were Curtiss planes. Daugherty estimated he invested more than $50,000 in his airfield and airplane services.

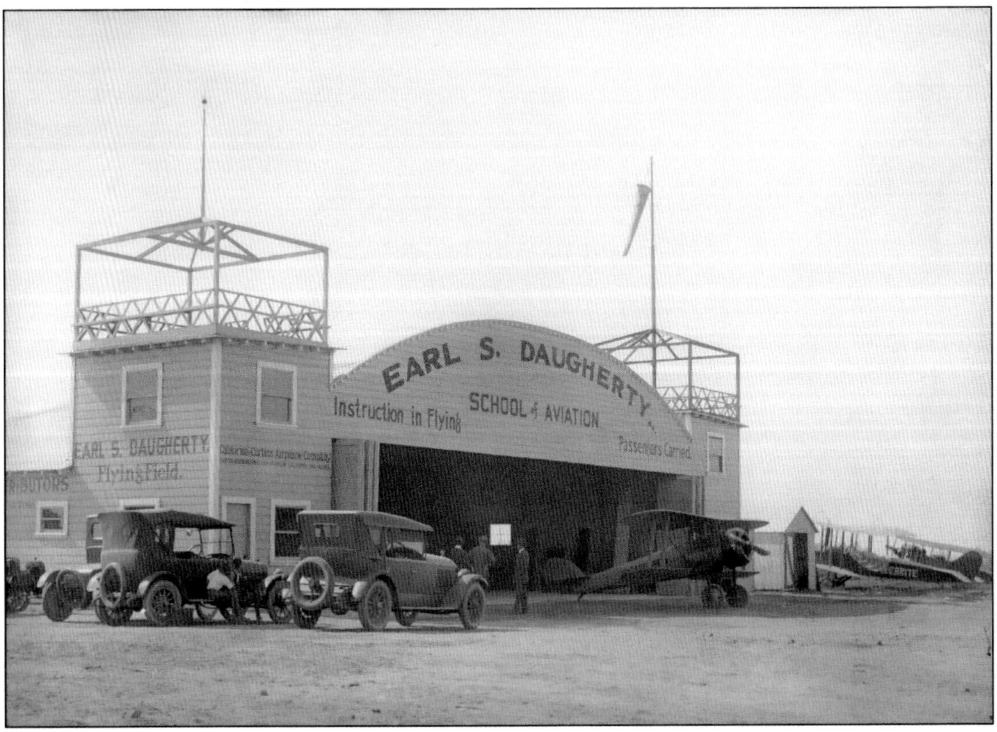

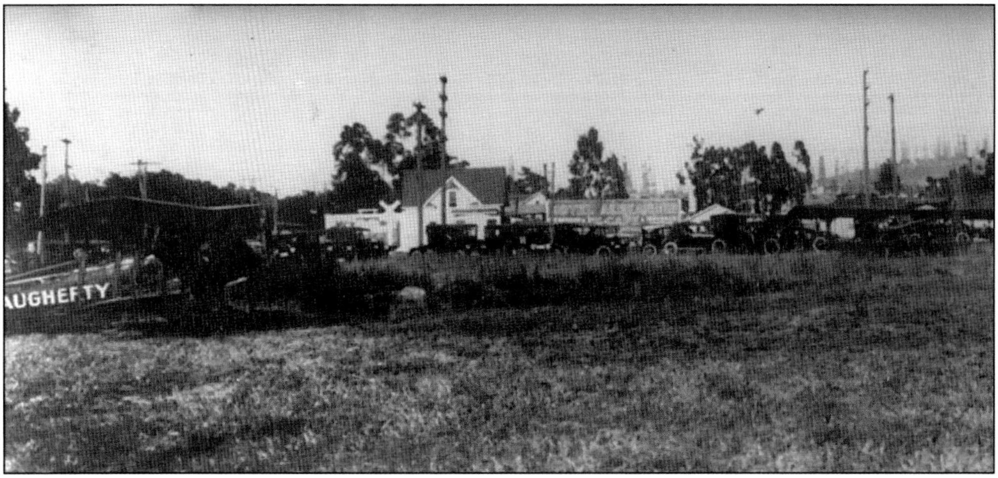

EARHART COMES TO LONG BEACH. Amelia Earhart and her father attended Earl Daugherty's air circus on December 28, 1920, at his field (pictured above and below). The 23-year-old Earhart moved to the Los Angeles area to be with her parents Edwin and Amy. She had seen airplanes in the Midwest, but Daugherty's flying skills were inspiring and the planes more sophisticated in Long Beach. After viewing the stunts, Earhart asked to be taken up in a plane. The next day, Long Beach Poly High School graduate and former Army Air Corps pilot and instructor Frank Hawks gave Earhart her first ride in an airplane at the Rodgers Airport on Wilshire Avenue, forever changing her life and aviation history.

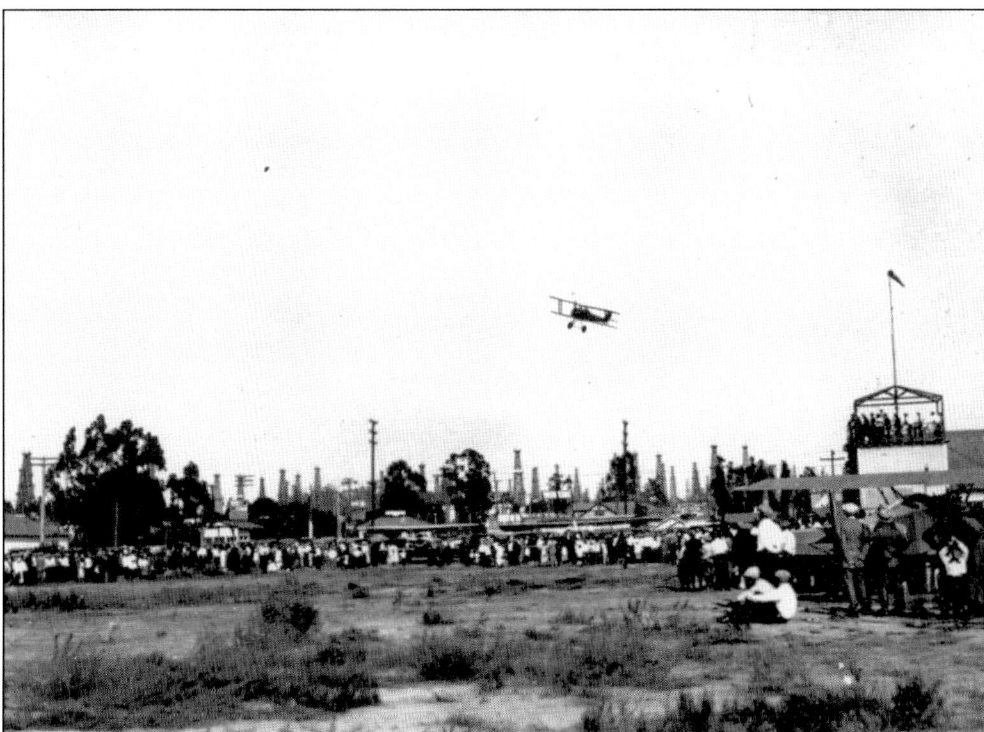

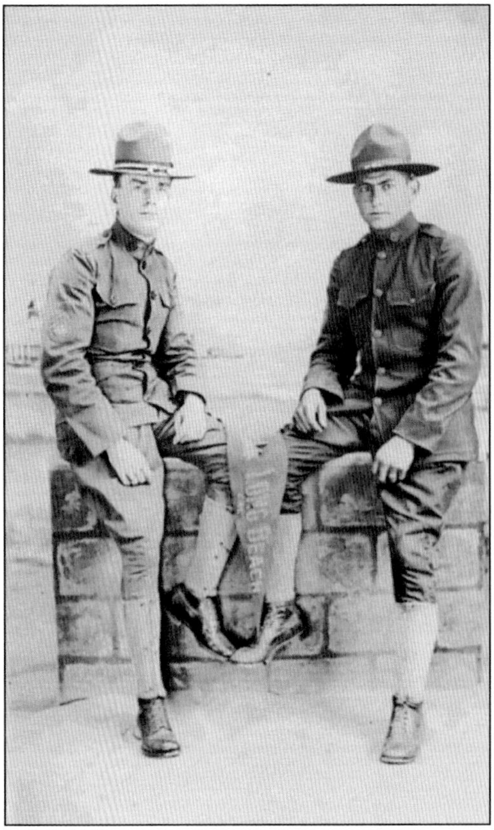

JOHN G. "MONTY" MONTIJO. Earl Daugherty and Frank Hawks were not the only Long Beach area aviators to have an impact upon the career of Amelia Earhart. John G. "Monty" Montijo (right), a former Army Air Corps instructor, came to Long Beach after being discharged and taught Earhart how to fly solo and to perform complicated aerobatics. Montijo is possibly the first licensed Hispanic flyer in the United States.

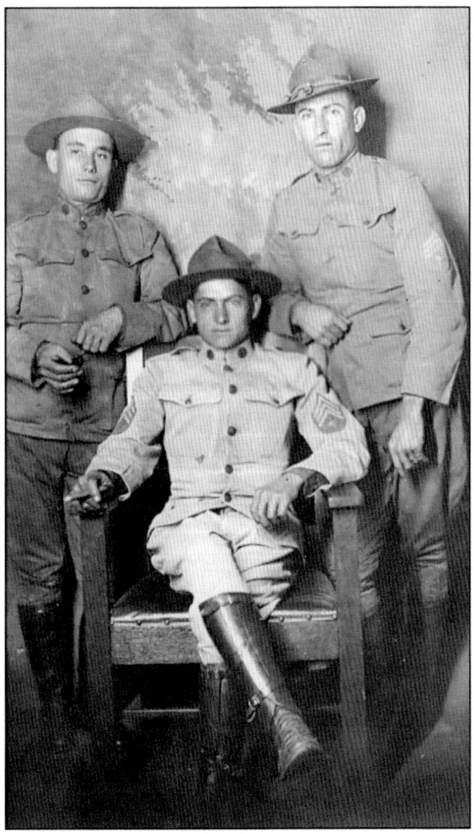

FLIGHT INSTRUCTOR. Montijo (seated), born in Jenny Lind in Calaveras County, California, enlisted in the army in 1916, learned how to fly, and became a test pilot and an instructor. He was a master sergeant/master electrician, the highest level of non-commissioned officers. Daugherty and Montijo served at Rockwell Field and March Field and taught over 1,000 students, giving each four weeks of training that included spirals and ordinary maneuvers, smooth landings, and cross-country flying.

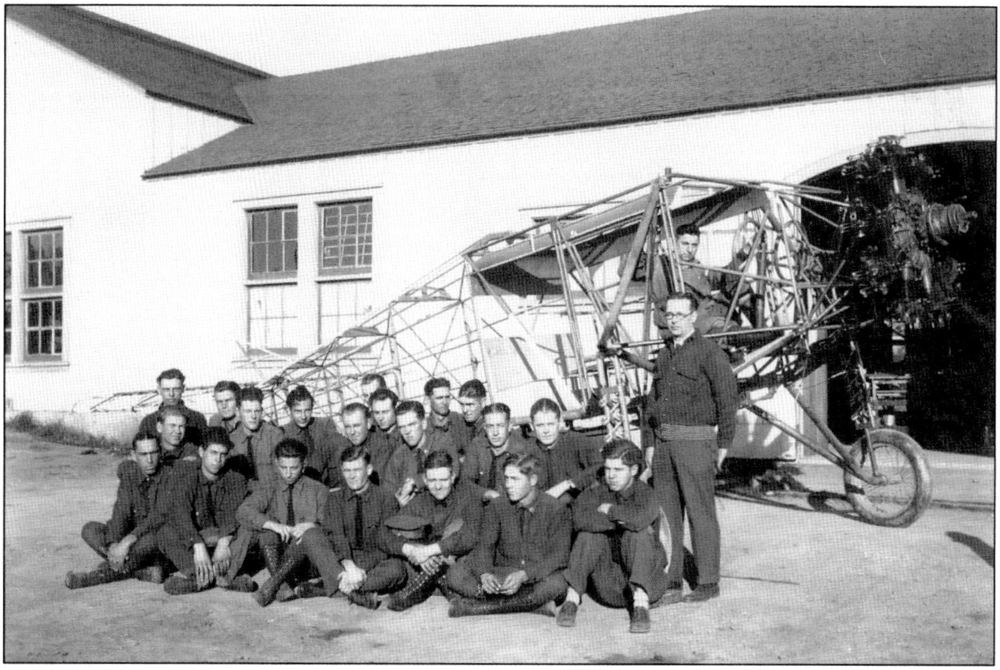

WILLIAM RANDOLPH HEARST'S PILOT. During his service, Montijo was credited for perfecting the "outside tail spin," which pilots used to evade the enemy. After the service, Montijo and his wife, Alta, traveled to Lima, Peru, where he oversaw an airfield built by U.S. promoters. Montijo returned and performed stunts for Goldwyn Studios, performing the first stunt of picking a man out of an automobile while flying overhead. He also served briefly as William Randolph Hearst's private pilot and taught aeronautics at California State at San Luis Obispo (above, inside frame of plane) where he and students built the *Glenmont* (below).

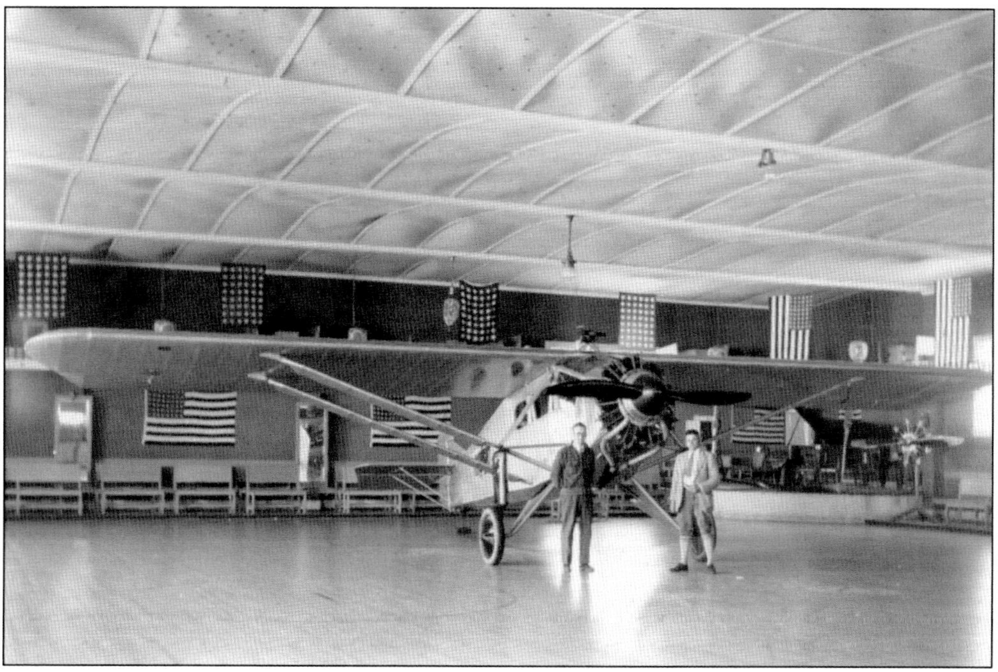

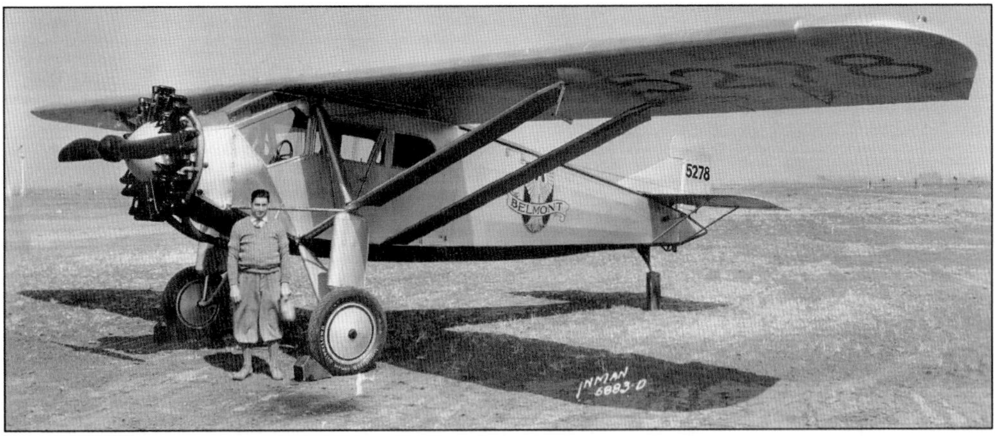

***Glenmont* Becomes *Belmont*.** John Montijo brought the *Glenmont* to Long Beach and sold it to C. B. "Boone" Bellows, a local Dodge Brothers dealer who renamed the plane *Belmont*. Montijo worked briefly for Bellows as a car salesman. The press reported that the *Belmont* was a "landau sedan six place semi-cantilever cabin monoplane. . . . the plane has the quickest rise and slowest landing speed for one of its size and horsepower with a 260 horsepower Salmson motor. . . . it maintains a cruising speed of 85 mph . . . remarkable inherent stability, no stabilizer being required and is said to fly hands off at 28 mph with full load. . . . landing speed is 30 mph . . . lands at 25 mph. . . . it is constructed of mahogany and plywood and with dual control it flies like an automobile."

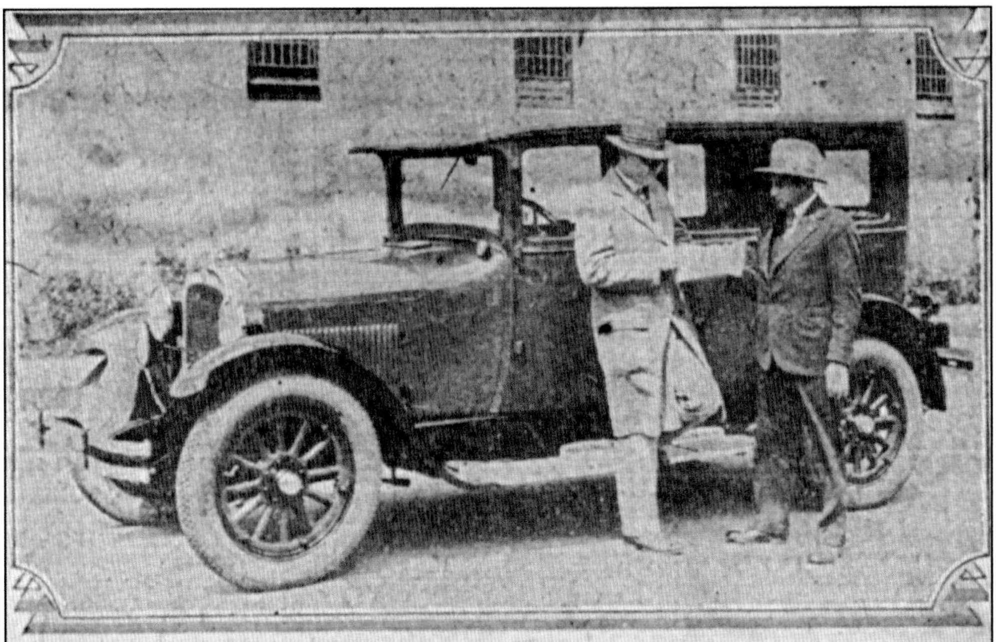

CHOICE MADE BY BOND SALESMAN.

Having Covered 34,000 Miles in Eighteen Months in His Third Dodge Brothers Car, With Remarkable Freedom From Mechanical Troubles and After the Genuine, Heavy Plate Glass of the Sedan Stopped a Stray Bullet, Thus Saving the Life of an Occupant, E. E. Morrison, Bond Salesman, Chose a New Dodge Brothers Special Sedan as the Car Best Suited to His Transportation Needs. J. G. Montijo, Salesman for C. B. Bellows, local Dodge Brothers Dealer, Is Shown at the Right, Delivering the Car and Bill of Sale to Mr. Morrison.

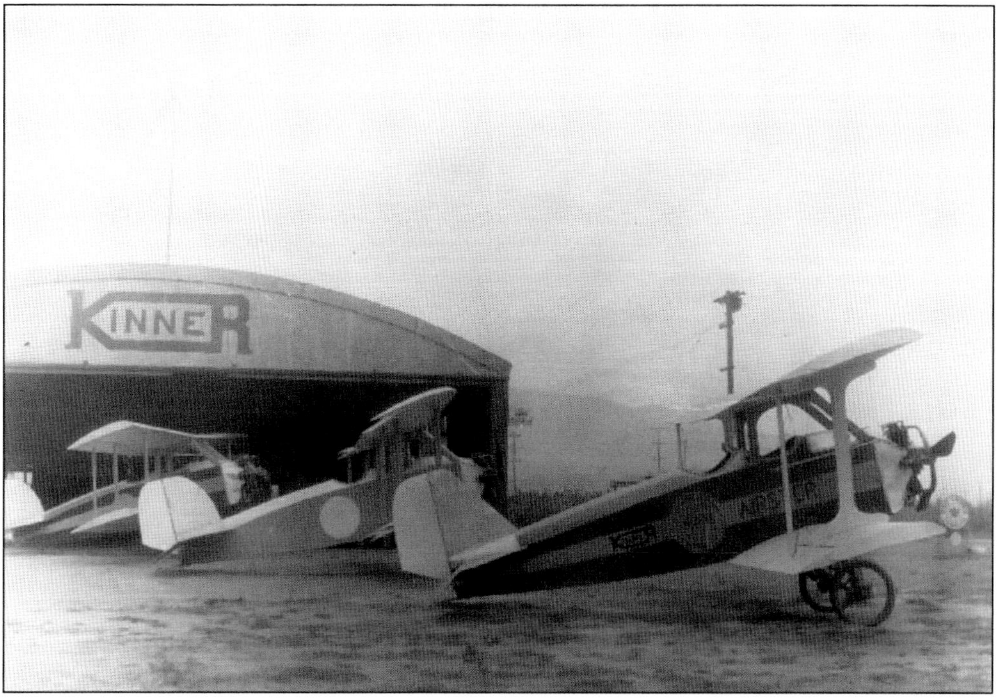

KINNER FIELD. Montijo tested airplanes at and around Long Beach and taught aerobatics. He was in business with W. B. Bert Kinner (a Long Beach resident), who owned and operated Kinner Field (above) on Long Beach Boulevard and Tweedy Road. Kinner's Airplane and Motor Corporation produced the Kinner Airster, which Amelia Earhart (right) purchased to take flying lessons. Earhart took a few flying lessons from Neta Snook, who worked at Kinner and was the first female graduate of the Curtiss School of Aviation. Alta Montijo defended John Montijo as being the person who taught Earhart to fly when she wrote a letter to the editor about Snook and Earhart parting ways for a reason "which was only known by Nita and John." Montijo finished Earhart's lessons until she could solo in her in a small, two-place biplane with a three-cylinder, 60-horsepower, air-cooled engine. Neta Snook quit flying. (Above, GLSPC; right, LOC.)

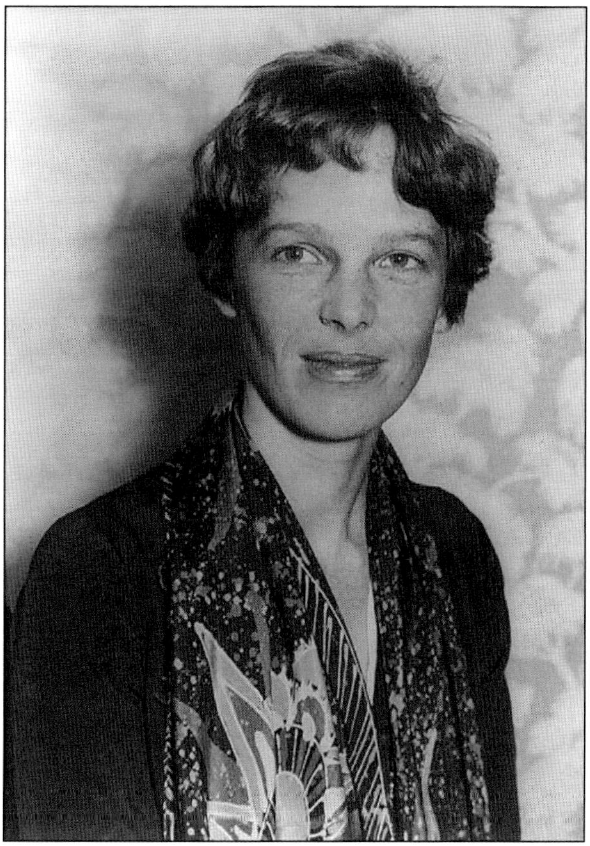

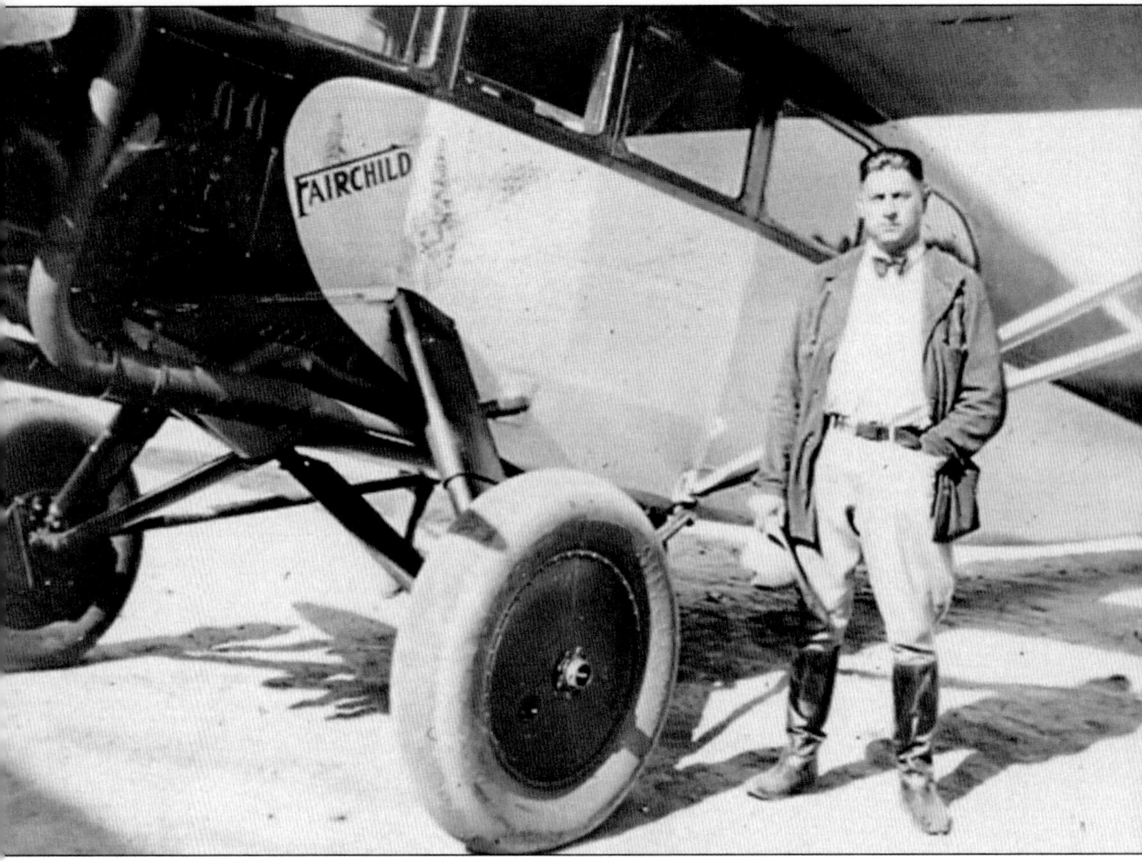

Montijo Recalls Earhart. Montijo (pictured) recalled his experience with Earhart in a 1935 *Christian Science Monitor* article: "Amelia Earhart was one of 3 girls who brought their planes to me in 1922 at Kinner Airport near Los Angeles and asked for instructions. The other two girls were an Osage Indian girl and Miss Alosia McClintock[,] a close fried of Amelia Earhart. Each girl had purchased an airplane and wanted instructions. . . . Amelia Earhart was an excellent student and was very thorough in what she undertook. Her first solo was with a perfect takeoff and after 10 minutes she made a landing worthy of a veteran. She never exhibited a bit of nervousness, even when she went solo after her 7th lesson." The article noted that Earhart was dressed in "khaki breeches and always carried a book which she read in spare moments" and "she walked 6 miles each day from Los Angeles to Kinner Airport" for her lessons.

OFFICIAL PROGRAM

AIR RODEO SATURDAY, MAR. 17, 1923 Glendale Airport

Under the Auspices of the Commercial Aircraft Association **Sanctioned by the Aero Club of Southern California**

Opening Event. Grand Formation

Music Furnished by Glendale Elks Band

1. Stunting Exhibition.
   Earl S. Daugherty with Canadian Curtiss.
2. Ladies Sportplane Special.
   Miss Amelia Earhart flying Kinner Airster
   Miss Andree Peyre flying Sport Farman
3. Accurate Landing Contest.
   Contestant's mark is T in center of field.
   (a) From altitude of 500 ft. motors idling
   (b) From altitude of 2000 ft. motor dead
   (Entrys to be announced)
4. Balloon Strafing.
   Airplane in mimic war combat
   Catron & Fisk Sportplane piloted by Frank Clarke
5. Parachute Jump.
   Miss Gladys Roy will jump 2000 ft from plane.

Exhibits Open to Public Inspection
S-E-5 Airplane used by Capt. Tait-Cox in his sky-writing exhibition.
THE ARROW. The giant 8-passenger transportation ship built by Rogers Aircraft Inc., Los Angeles, for the Airline Transportation Co. This ship will be placed in daily passenger service between Los Angeles and Imperial Valley points.
KINNER AIRSTER. A sport plane manufactured by the KinneR Airplane and Motor Corp., Glendale, Cal.
SPORT FARMAN. French sport Plane owned by Miss Andree Peyre
GERMAN FOKKER with 150 h.p. Hisso owned by Victor H. Staheli.
ITALIAN BALLILA 220 h.p. flown by Frank Clarke.
FRENCH SPAD VII 180 h.p. Hisso owned and flown by Earl S. Daugherty.
LA PERE 400 h.p. Liberty owned by L. C. Brand.

6. Skywriting Exhibition.
   Capt. Leslie R. Tait-Cox, master aerial penman, flying S-E-5 airplane at 125 m.p.h. will write "Lucky Strike" 10,000 ft. above the Glendale Airport.
7. Jenny Scramble.
   A 5-mile open event for Curtiss J. N. 4 D's (Entrys will be announced 10 minutes before Race.)
8. Bullet Drop.
   Price Miller will jump from airplane, turn parachute inside out, and fall to within 500 ft. of the ground before opening again.
9. St. Patrick's Day Special.
   A 10-mile race of So. California's fastest airplanes.
10. Jenney Relay Race.
    3 laps. Entries to be announced.

The Commercial Aircraft Assn. will furnish Richfield Gasoline exclusively to all airplanes at the field.

Added event. If weather conditions are suitable Miss Amelia Earhart, flying her Kinner Airster, will attempt to break the altitude record of 13,200 ft. for plane equipped with 50 H.P. motor.

The Commercial Aircraft Assn. will furnish Pennzoil Aviation Oil exclusively to all airplanes at the field.

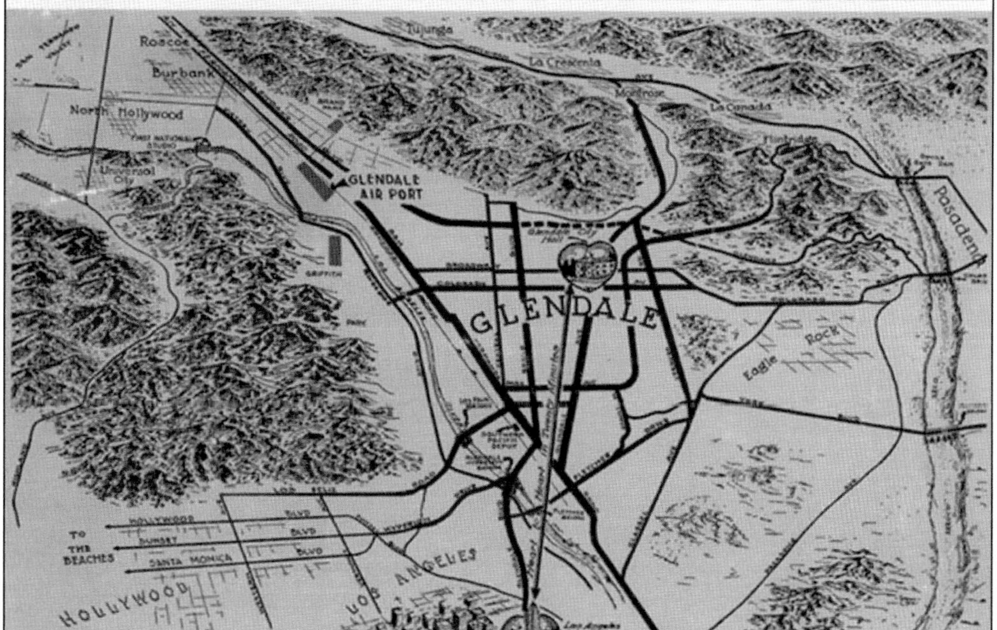

HELPING GLENDALE. After Montijo certified Earhart and her friend Alosia McClintock in advanced aeronautics and aerobatics, they put on a "fine exhibition of aerial aerobatics over Pasadena" and then performed in the Long Beach Air Rodeo with Daugherty, Frank Clarke, Frank Hawks, and Wesley May. Cities, like Glendale, scrambled to establish municipal airports with the announcement of a U.S. Air Mail service. Daugherty, who built planes with Billy Meyer in Glendale, worked to encourage voters to approve an airport bond measure with an air rodeo on March 17, 1923. Daugherty performed a "stunt exhibition" in both a Canadian Curtiss and a French SPAD VII Hisso. Earhart performed in her Kinner Airster. The event was a success, but the bond failed, and Glendale was unable to establish a municipal airport for many years. (GLSPC.)

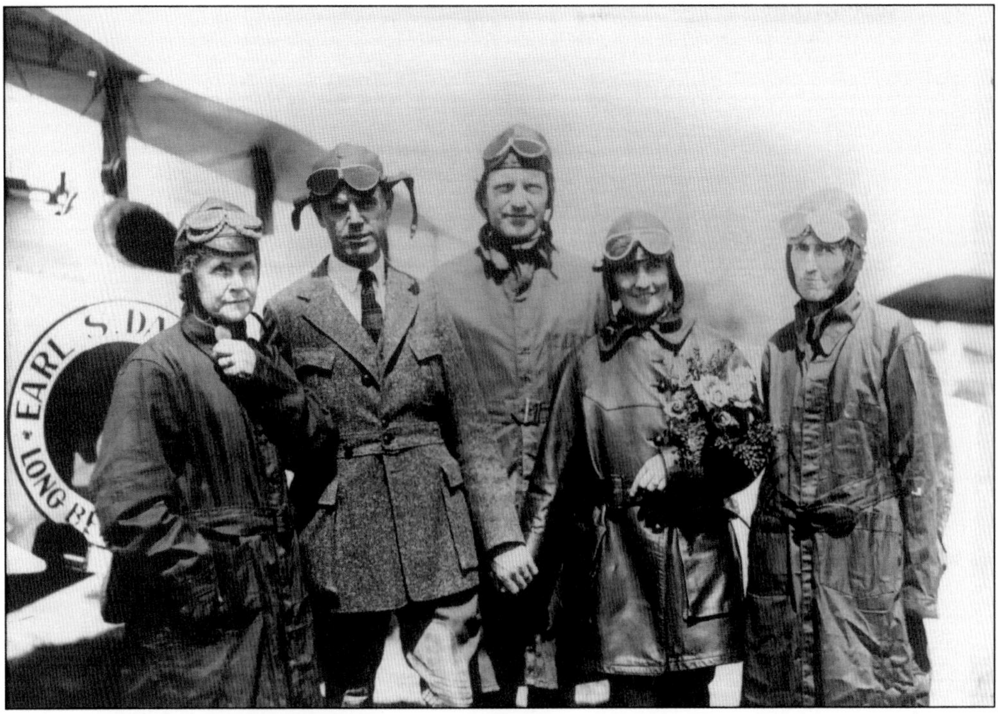

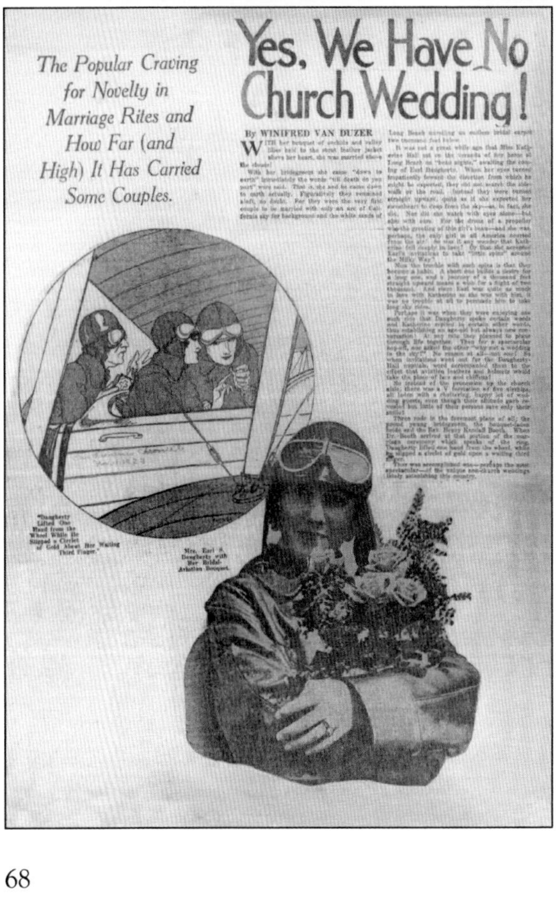

The Popular Craving for Novelty in Marriage Rites and How Far (and High) It Has Carried Some Couples.

Yes, We Have No Church Wedding!

By WINIFRED VAN DUZER

"Daugherty Lifted One Hand from the Wheel While He Slipped a Circlet of Gold About Her Waiting Third Finger."

Mrs. Earl S. Daugherty with Her Bridal-Aviation Bouquet.

DAUGHERTY WEDS. On September 7, 1923, Daugherty became the first pilot married while flying. Daugherty piloted a wood-and-fabric Orenco Tourister, designed by the Ordnance Engineering Corporation and produced by Curtiss Aeroplane and Motor Company. He married Catherine "Kay" Hall in a ceremony performed by First Congregational Church pastor Rev. Henry K. Booth, while flying 2,000 feet over "the white sands of Long Beach." Daugherty's parents and Kay's mother accompanied them. Following behind in a V-formation were several airships and their passengers. The Daughertys honeymooned in Coronado.

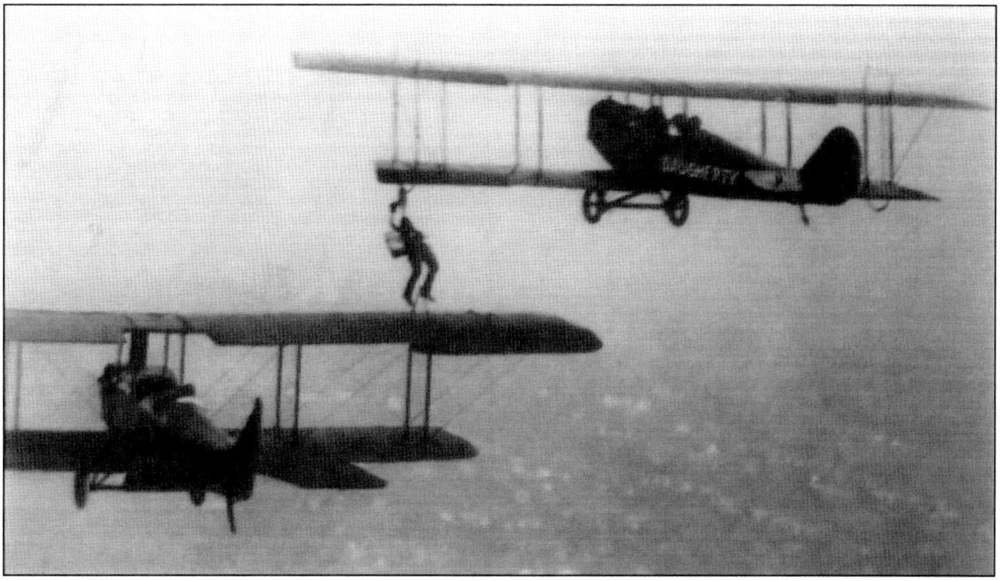

Midair Fueling. During an air meet on November 12, 1921, Frank Hawks and Earl Daugherty (pictured) performed the first ever midair refueling. Before the stunt, Wesley May and Freddie Lund "walked" between airplanes in the air. May climbed from Hawks's plane onto Daugherty's with a 5-gallon can of gasoline strapped to his back and then "replenished an imaginary failing supply" of fuel. Then he stood on his head on the plane's upper wing. Lund did a similar stunt between the planes piloted by Frank Tomick and Frank Clarke. The planes came so close together that the wings nearly touched. (Lund performed stunts for a number of film companies, as shown in the picture. Montijo served in the army with Frank Tomick, and both Tomick and Clarke later worked for Howard Hughes during the filming of *Hell's Angels.*) Midflight refueling became a critical maneuver for military aircraft in World War II. The 1921 event is featured as a painting in the Leach International Heritage of the Air Collection of key events in aviation history. (Right, LOC.)

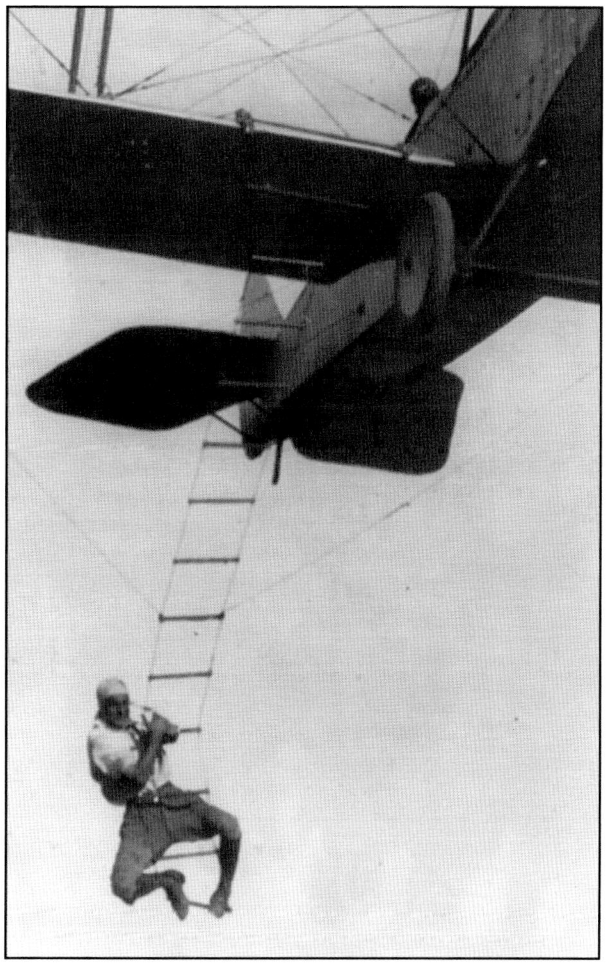

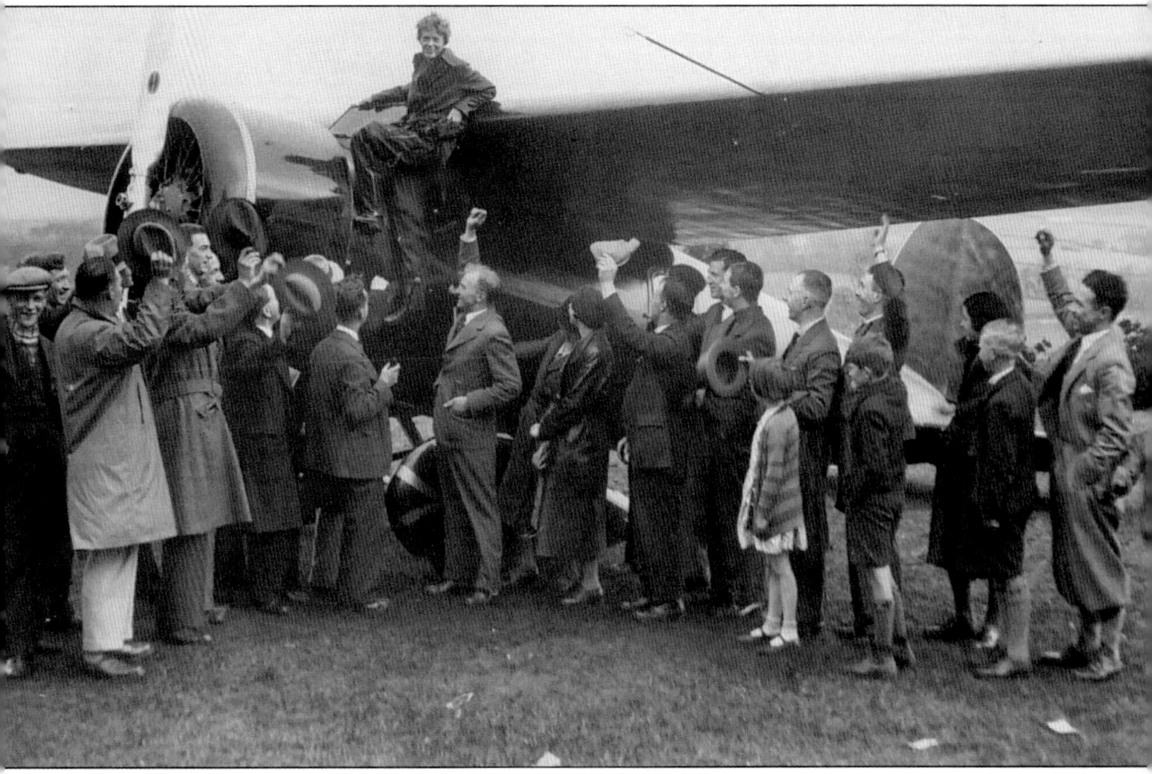

EARHART LANDS IN LONDONDERRY. Exactly five years after Charles A. Lindbergh crossed the Atlantic, Montijo's student Amelia Earhart became the first woman to fly the Atlantic solo when she flew her Lockheed Vega from Harbor Grace, Newfoundland, and landed in Londonderry, Ireland (pictured). Her flight took 14 hours and 54 minutes and covered 2,026 miles. Earhart was honored by several European countries and presented the National Geographic Gold Medal in aviation science by Pres. Herbert Hoover. By Joint Resolution of the House and Senate, Earhart was awarded the Distinguished Flying Cross for "displaying heroic courage and skill as a navigator, at risk of her life, by her nonstop flight in her plane, unnamed, from Harbor Grace, Newfoundland, to Londonderry, Ireland on May 20, 1932 by which she became the first and only woman, and the second person, to cross the Atlantic Ocean in a plane in solo flight, and also establish new records for speed and elapsed time between the two continents." (AC.)

Earhart's Reflection on Flight. Montijo was one of the first people to write Earhart after her historic flight in 1932. In return, she him a letter (pictured), thanking him for his letter of congratulations and reflecting on how she viewed her flight over the Atlantic as no more difficult than an overland flight.

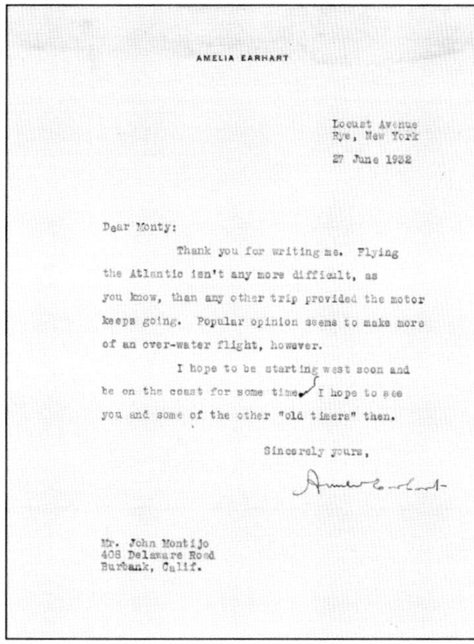

AMELIA EARHART

Locust Avenue
Rye, New York

27 June 1932

Dear Monty:

Thank you for writing me. Flying the Atlantic isn't any more difficult, as you know, than any other trip provided the motor keeps going. Popular opinion seems to make more of an over-water flight, however.

I hope to be starting west soon and be on the coast for some time. I hope to see you and some of the other "old timers" then.

Sincerely yours,

Amelia Earhart

Mr. John Montijo
408 Delaware Road
Burbank, Calif.

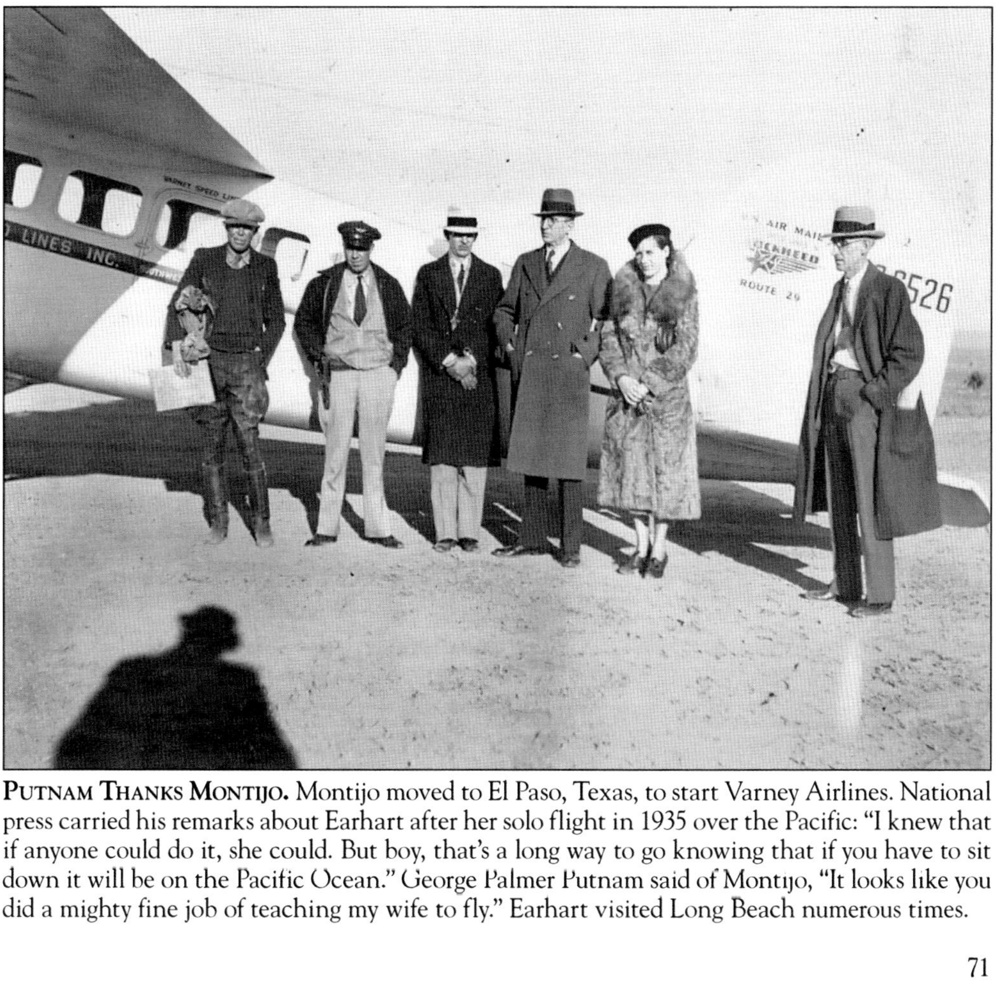

Putnam Thanks Montijo. Montijo moved to El Paso, Texas, to start Varney Airlines. National press carried his remarks about Earhart after her solo flight in 1935 over the Pacific: "I knew that if anyone could do it, she could. But boy, that's a long way to go knowing that if you have to sit down it will be on the Pacific Ocean." George Palmer Putnam said of Montijo, "It looks like you did a mighty fine job of teaching my wife to fly." Earhart visited Long Beach numerous times.

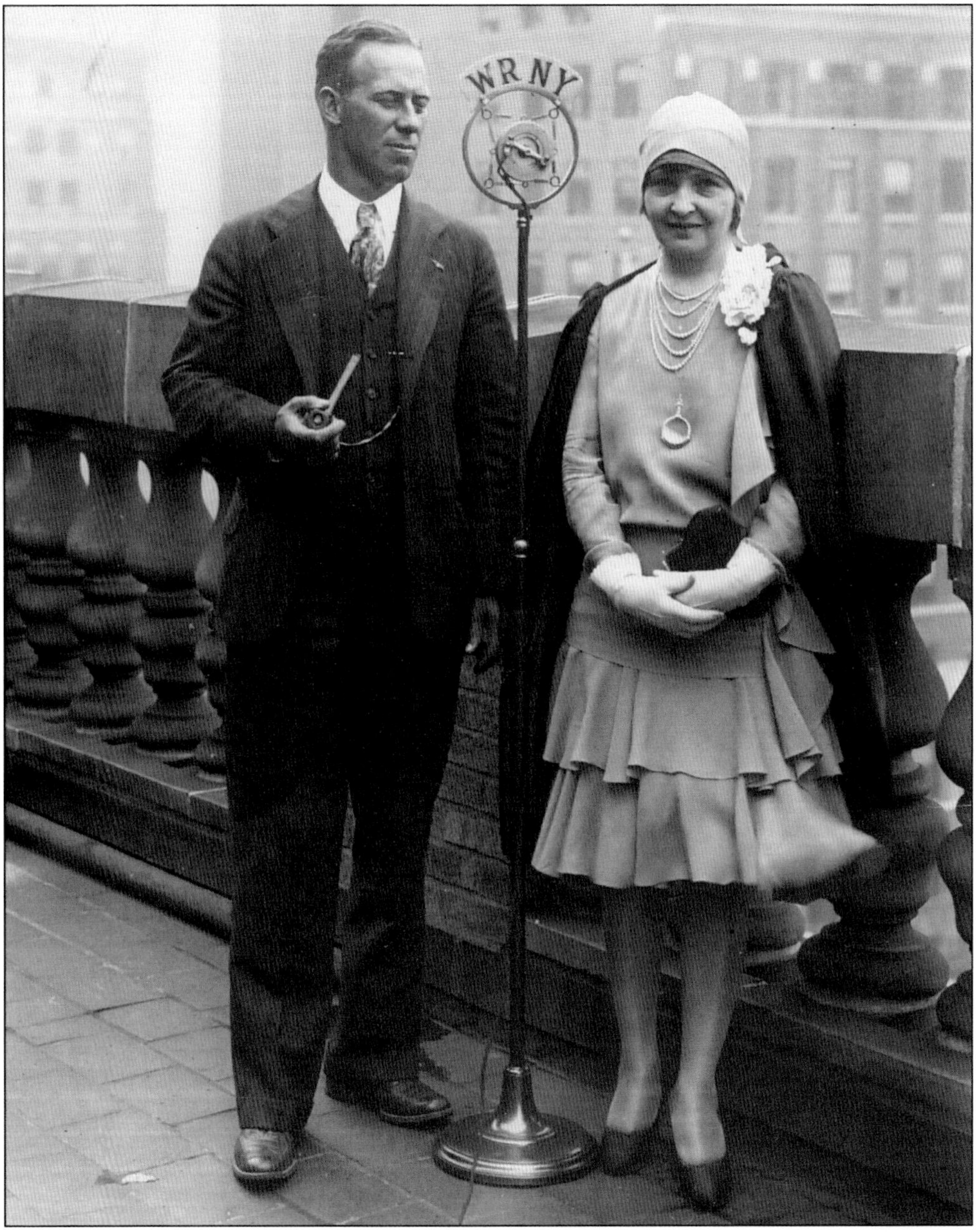

CAPTAIN HAWKS. The man who took Earhart up in her first plane ride set a few records of his own. Capt. Frank Hawks is shown with his wife, telling radio listeners the story of his record-breaking fete of flying a glider in June 1929 from California to New York and back in 44 hours. He broke his own record on the west-east journey, established a new record on the east-west leg, and set a transcontinental nonstop record of 18 hours and 21 minutes from California to New York in his Lockheed Vega. Hawks flew several years advertising Texaco oil products and was a "super hero" to kids belonging to the Frank Hawks Air Hawks Club, who wore their "wings" ordered from Post Bran Flakes. A weekly *Captain Hawks* comic was featured next to *Little Orphan Annie* in the newspapers. Hawks was killed in 1938 in the crash of his Gwinn Aircar. Earhart was lost at sea in July 1937. (AC.)

Six

The Long Beach Municipal Airport

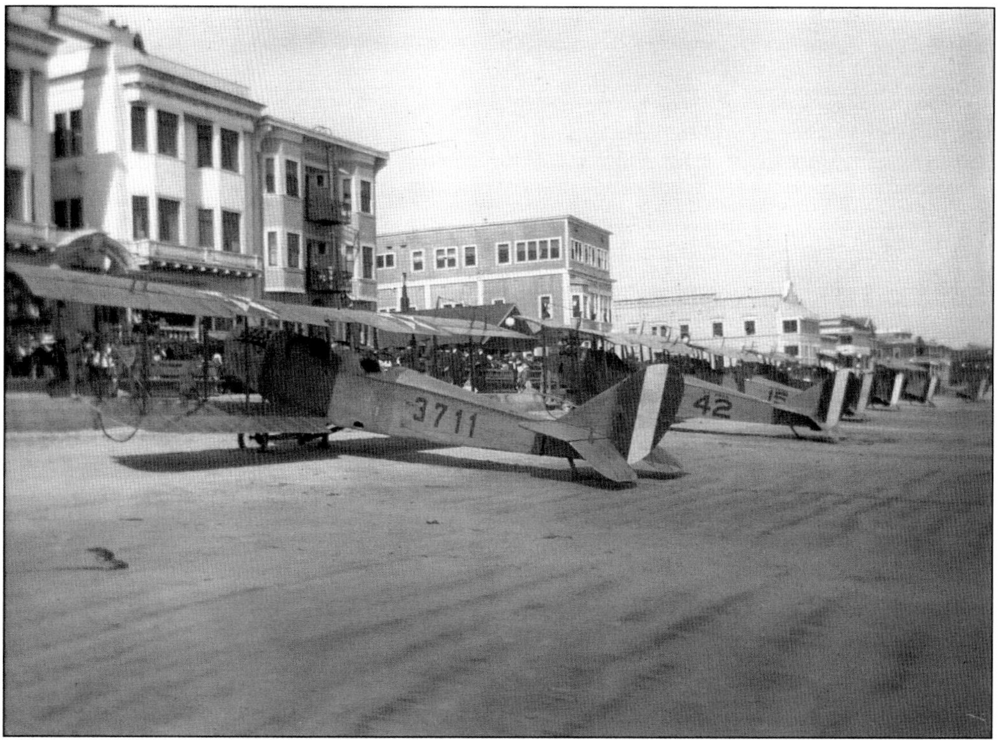

Long Beach Municipal Airport. Realizing that Long Beach could no longer accommodate aviation on its beach nor on Daugherty's small airfield inland, the city council in November 1923 dedicated 80 acres of water department land at Cherry Avenue and Spring Street (north of Signal Hill) for the Long Beach Municipal Airport, making Long Beach the first city in California to establish a municipal airport.

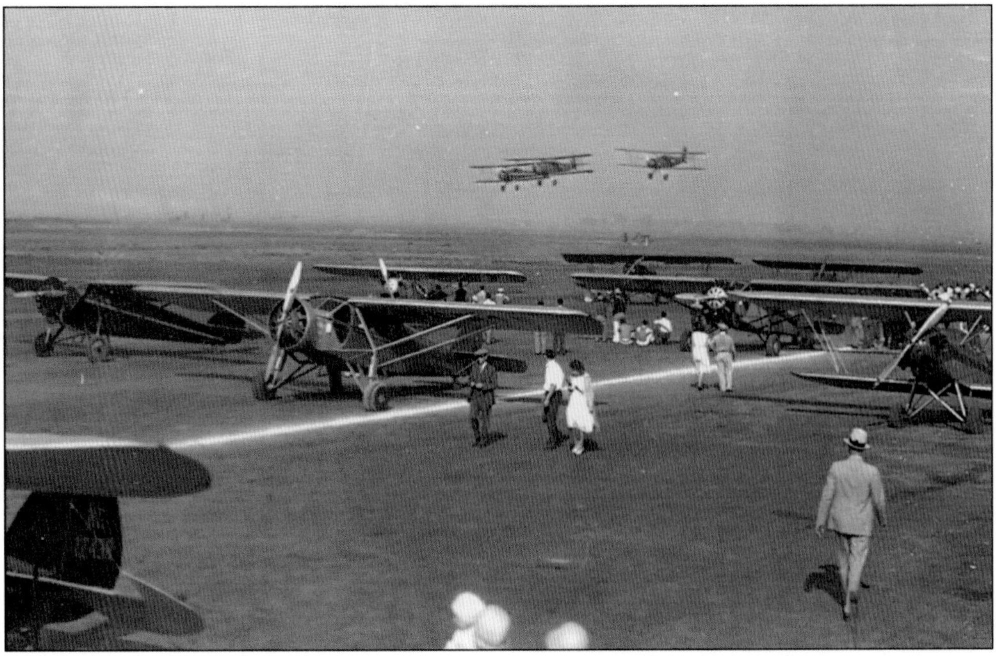

Airport Interest. Earl Daugherty's connections with the military brought the latest advances in aviation to the new Long Beach field and drew crowds at all the events even though the municipal airport and field were miles away from local residents. The Aerospace Industries Association of America 1922 Aircraft Yearbook acknowledged Daugherty's vision: "Earl S. Daugherty, at Long Beach, Cal., has shown it to be possible to carry on transportation and exhibition work by defining and strictly separating them, with the result that safety and interest are combined."

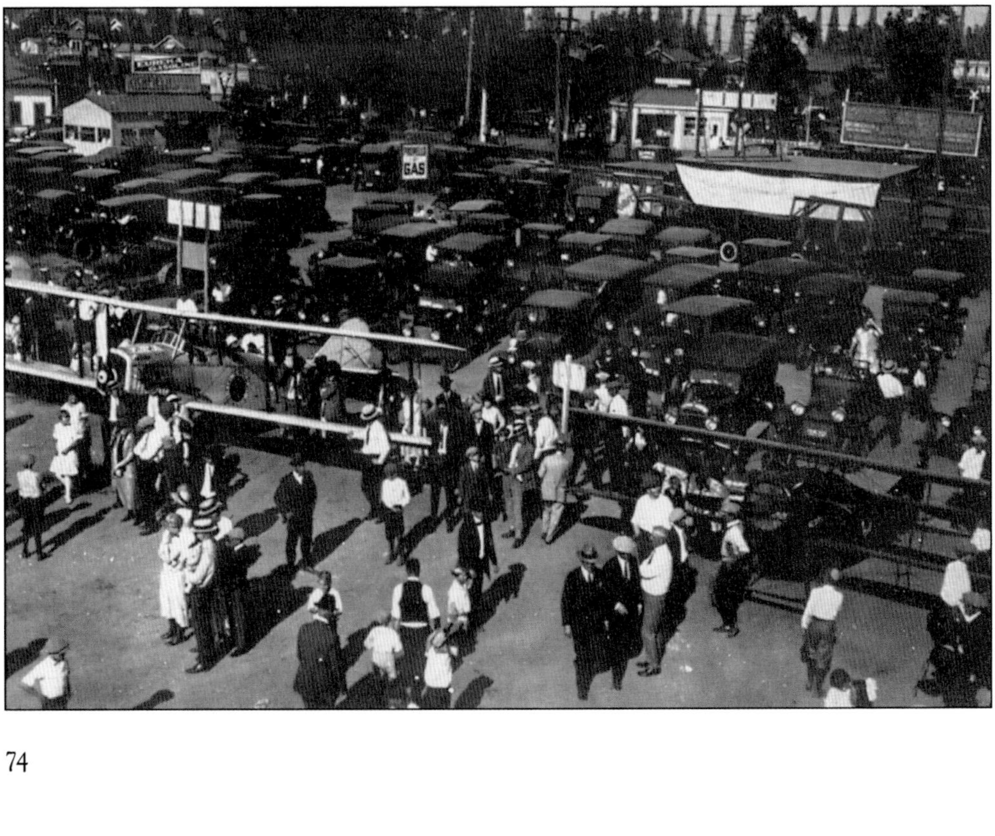

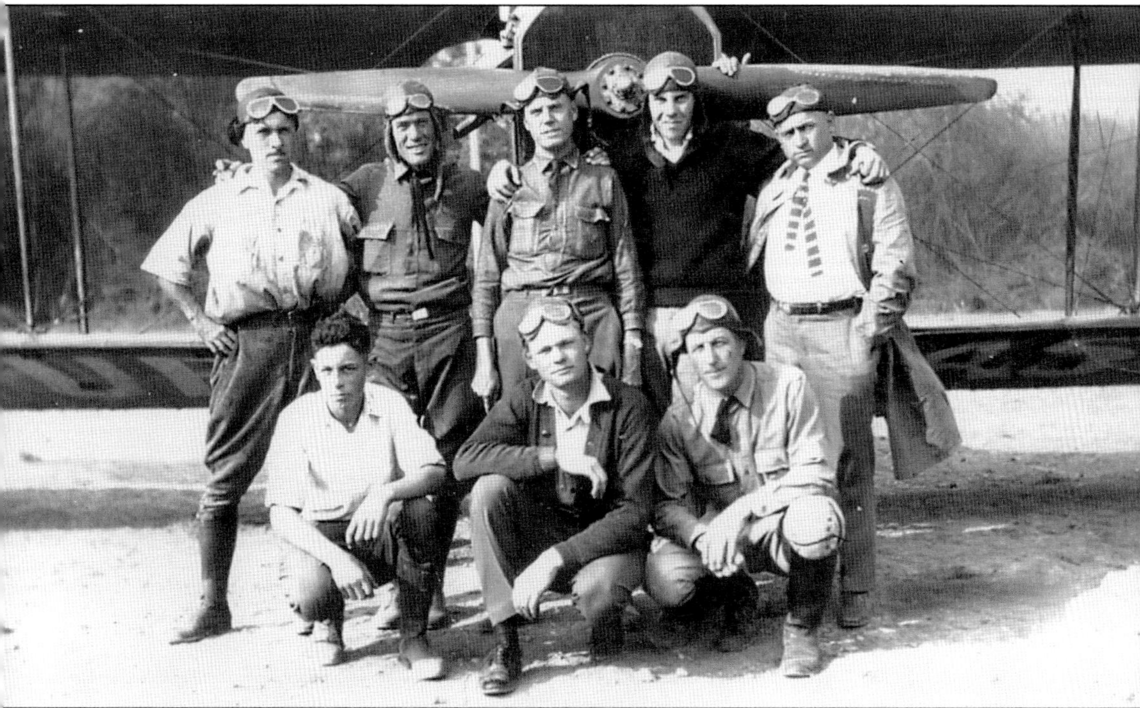

Long Beach Aero Club. Local Aero Clubs under the auspices of the Aero Club of America (a subsidiary of the Automobile Club) sponsored aviation meets, usually at fairground racetracks, and eventually advocated for a chain of "airways" across the country in the same manner the Automobile Club advocated for roads. Daugherty formed the Aero Club of Long Beach, which sponsored "air tournaments" and "air rodeos" at Daugherty Field. Pictured are J. G. Montijo, commodore (standing on the right, wearing a striped tie); Earl S. Daugherty, vice commodore (standing center); A. R. Ebrite, rear commodore, who leased the first hanger at the Long Beach Municipal Airport; J. G. York, secretary/treasurer; Watson Kurl, assistant secretary; and E. N. Ebrite, field marshal. The Aero Club promoted the "air rodeo" on May 30 and 31, 1924, promising the local press, "This meet will be the biggest every held in the city and one of the largest every projected in California. Preparations are being made by the Long Beach Aero Club for handling of 40,000 people."

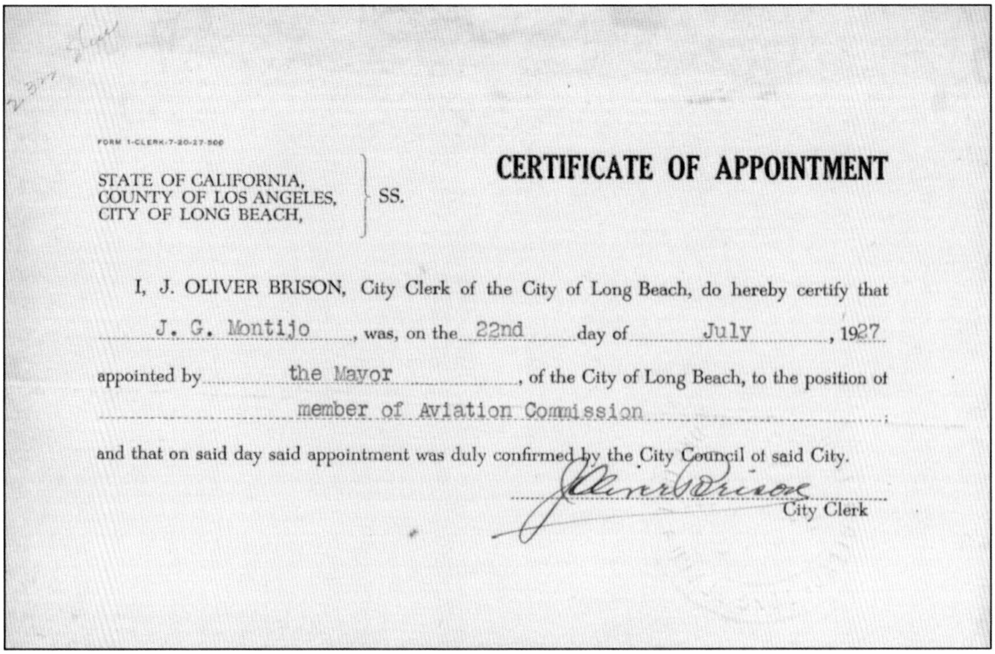

FORM 1-CLERK-7-20-27-500

STATE OF CALIFORNIA,
COUNTY OF LOS ANGELES, } SS.
CITY OF LONG BEACH,

CERTIFICATE OF APPOINTMENT

I, J. OLIVER BRISON, City Clerk of the City of Long Beach, do hereby certify that J. G. Montijo, was, on the 22nd day of July, 1927 appointed by the Mayor, of the City of Long Beach, to the position of member of Aviation Commission; and that on said day said appointment was duly confirmed by the City Council of said City.

J. Oliver Brison
City Clerk

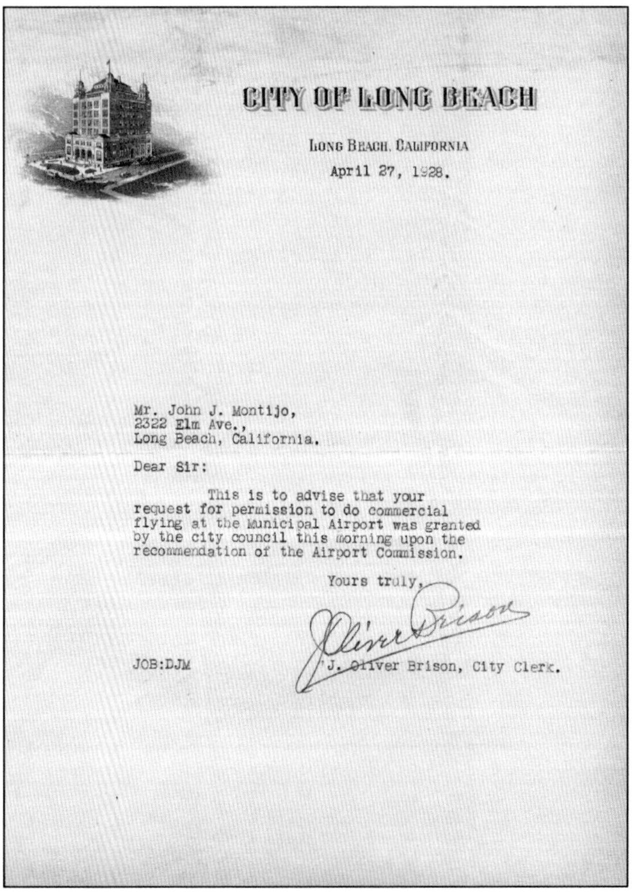

CITY OF LONG BEACH

LONG BEACH, CALIFORNIA
April 27, 1928.

Mr. John J. Montijo,
2322 Elm Ave.,
Long Beach, California.

Dear Sir:

This is to advise that your request for permission to do commercial flying at the Municipal Airport was granted by the city council this morning upon the recommendation of the Airport Commission.

Yours truly,

J. Oliver Brison, City Clerk.

JOB:DJM

AVIATION COMMISSION. The first Long Beach "aviation commission" was established in 1924, with Daugherty, Montijo, and A. R. Ebrite as commissioners. The city council decided who could fly out of the airport. Daugherty was removed from the commission in May 1928 by the recreation superintendent, D. D. Skinner, because he advocated weed removal and grading of the airfield and professional management of the airport. The news of his removal brought hundreds to the airfield to watch Daugherty fly. The Aeronautic Society protested, and the city attorney reinstated Daugherty. Montijo was reappointed in 1929. Ebrite operated the first civilian parachute school at the municipal airport. W. J. Putnam was named the first director of the Long Beach Municipal Airport, known then as the "Superintendent of Airports," in 1925. He served until 1940.

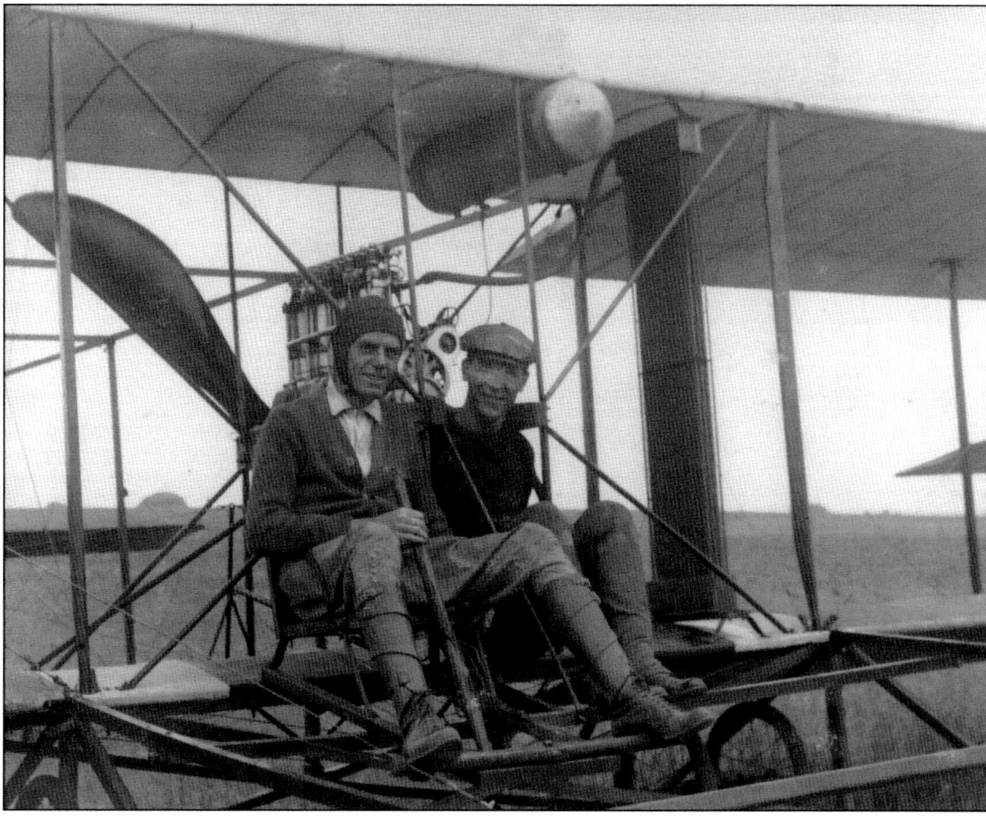

POLSON SPECIAL. Airplane design changed rapidly as is shown in the two photographs. The photograph above is of Daugherty carrying his first paying passenger. His first passenger was Nell Holman (née Shall), who accompanied him on the flight that qualified him for his license in 1911. Below is a photograph of several planes at the Daugherty Field in 1920 including the *Polson Special* (in foreground) that Daugherty and Bixby family chauffeur and mechanic Thor Polson built around 1915 at the Bixby Rancho. Daugherty flew the plane, winning first prize in the "Sports Plane Special" at the Aero Club of Southern California Air Meet at the Los Angeles Speedway. (Above, LBMA; below, Pat Loop.)

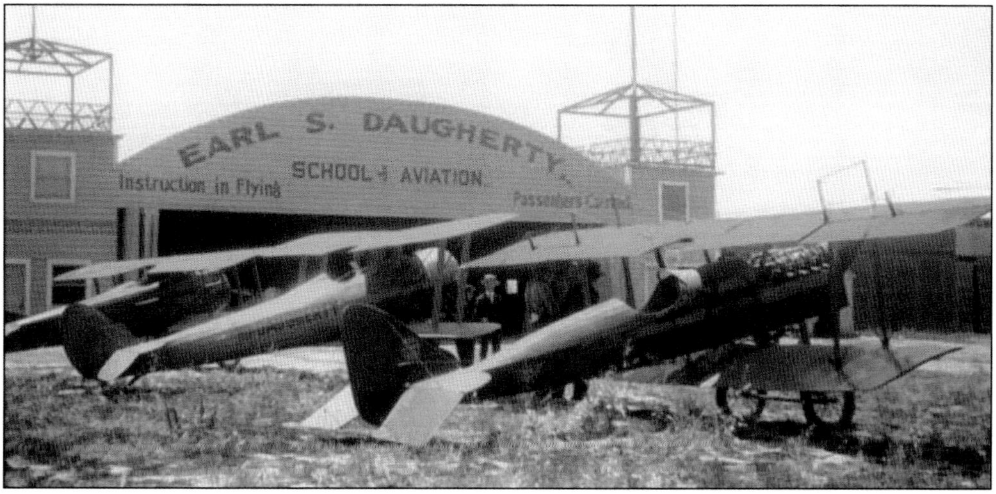

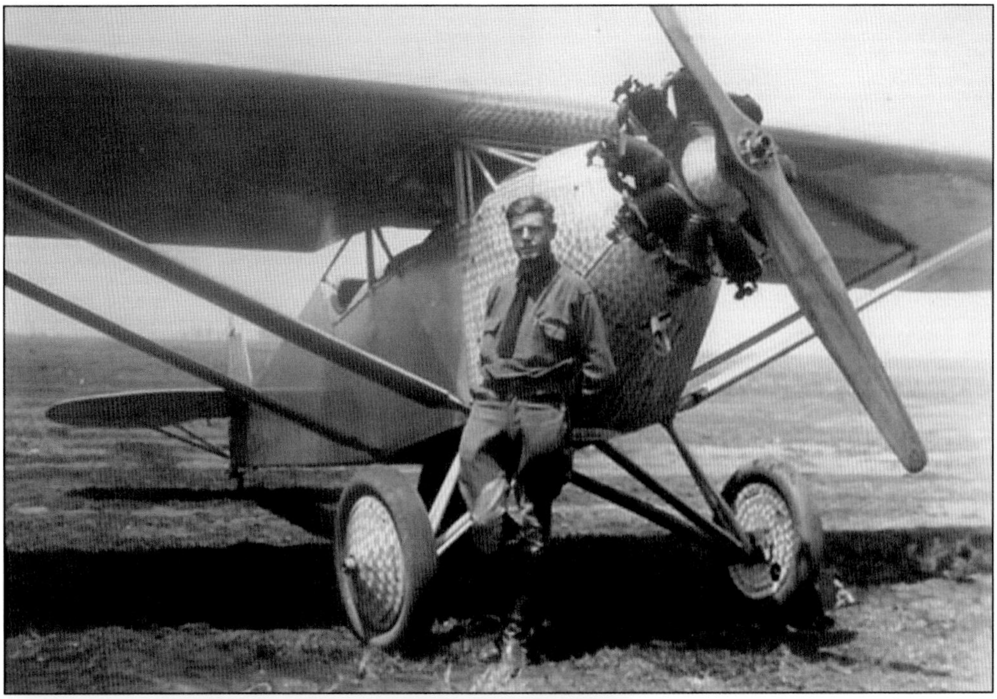

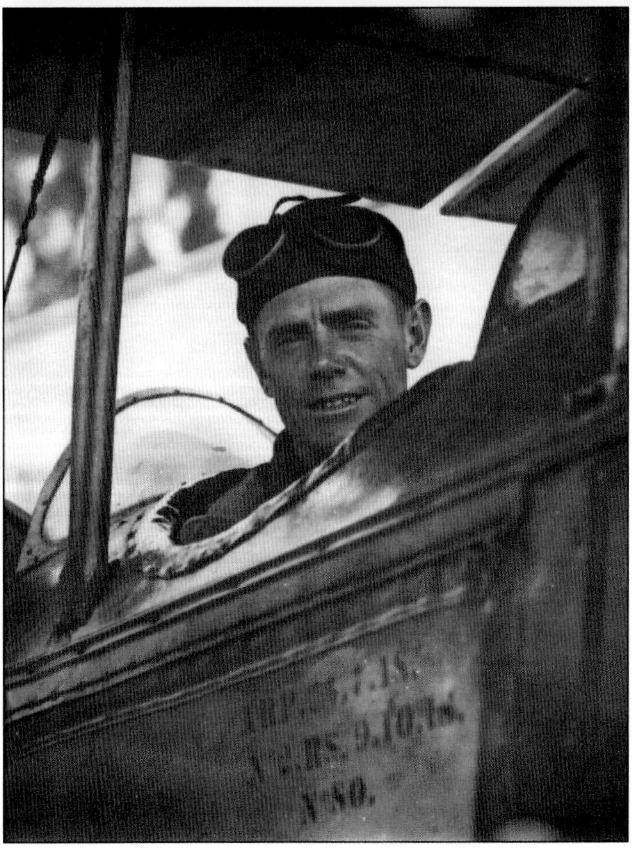

Passenger Record. The city's 1924 aviation ordinance prohibited stunt flying over inhabited sections inside municipal boundaries and prescribed certain "rules of air." By July 1925, Earl Daugherty set a record for carrying 7,000 passengers from the Long Beach Municipal Airport without a single mishap. However, October 1924, he was sued for $3,000 in damages because of a collision with his plane. Plaintiff John Anderson contended that Daugherty landed "with the wind not against it" and violated "aerial ethics." It was the first aviation-related lawsuit filed. Daugherty eventually closed his private field and focused on flight instruction, aerial photography, and special delivery, such as the $1,750,000 in bonds he delivered to Reno, Nevada. (Above, RTG.)

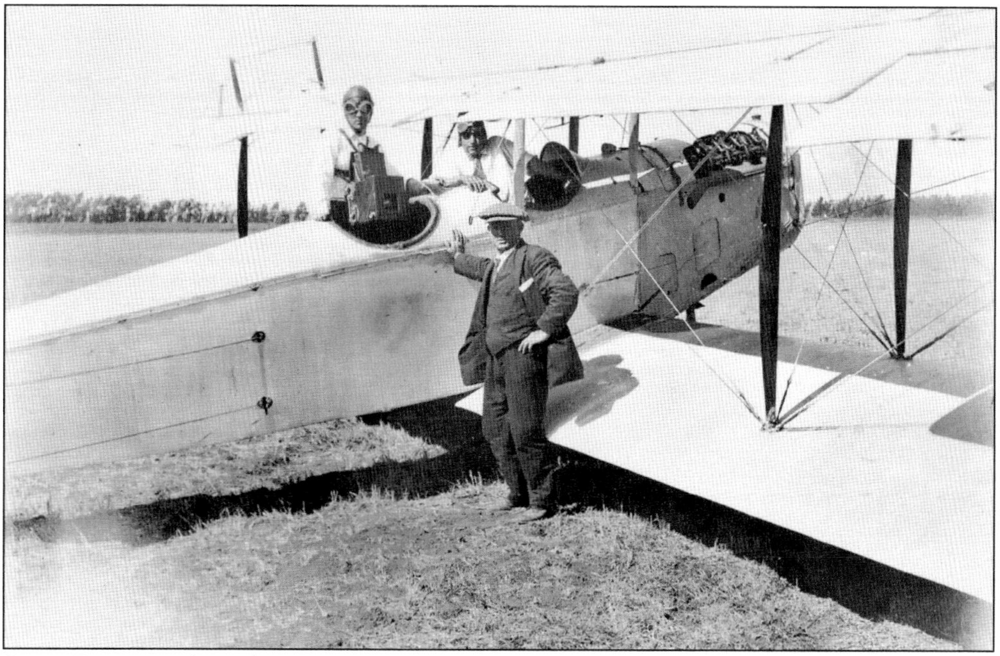

CABIN PLANE. Early airplanes were constructed of wood or bamboo and fabric. They were unstable and unpredictable. Open cockpits exposed pilots to the harsh elements and made conversations between pilot and passengers impossible, as can be seen in the photograph above of John Montijo in the pilot seat and a camera operator behind him. John Montijo designed and built the first "cabin plane" on the Pacific Coast. Earl Daugherty described his friend's plane to the press in 1924: "One of the newer ships will be J. G. Montijo's cabin cruiser, carrying five passengers besides the pilot. This is an entirely enclosed job with upholstered seats. It is the sedan of the air. Very little noise from the engine is heard inside the cabin. Passengers can converse in an ordinary tone of voice and need wear no special flying togs." The plane was destroyed in 1925 when a camera truck ran into it. Montijo sued the studios.

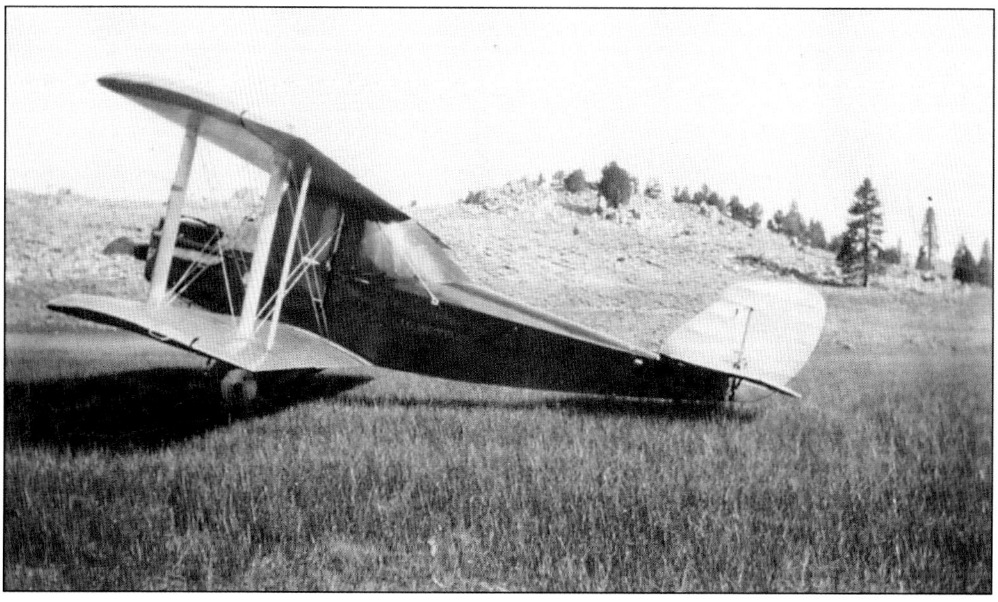

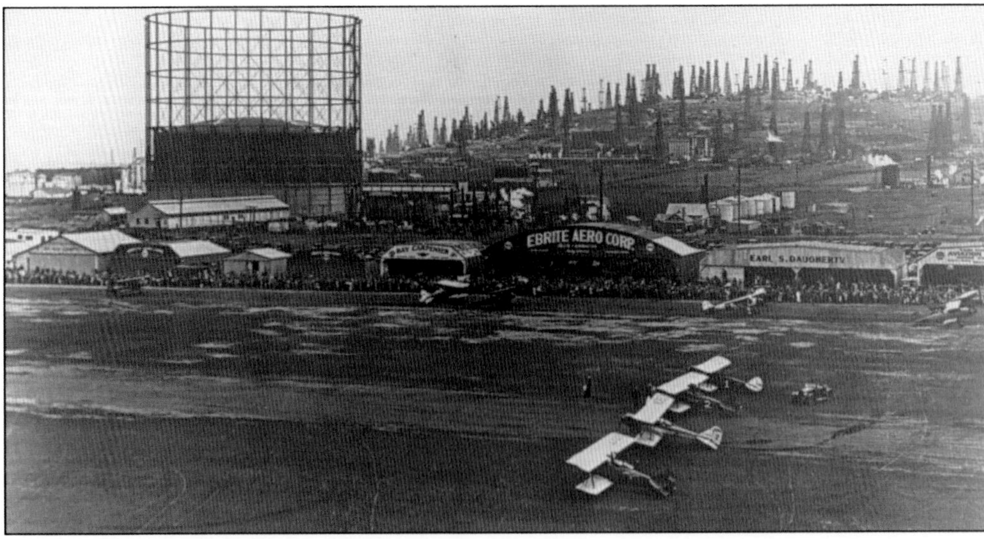

More Airport Tenants. Mechanical development of airplanes and engines after World War I made it possible for planes to fly more than 100 miles per hour and to carry pilot, passengers, and cargo or weapons. Aviators quickly moved from barnstorming, focusing instead on setting records for speed and distance and photo reconnaissance and mapping. Daugherty Field was also home to Continental Air Map Company, whose pilots did aerial photography for such clients as Standard Oil. The navy remarked after its pilots participated in a Daugherty air rodeo, "The Long Beach municipal aviation field at Spring Street and Cherry Avenue is one of the best in southern California. . . . the Long Beach field is remarkable, in that it is a four-way field, or a field where it is possible for aviators to land toward any quarter of the compass. It is possible to land on many fields in one way only." (Below, RTG.)

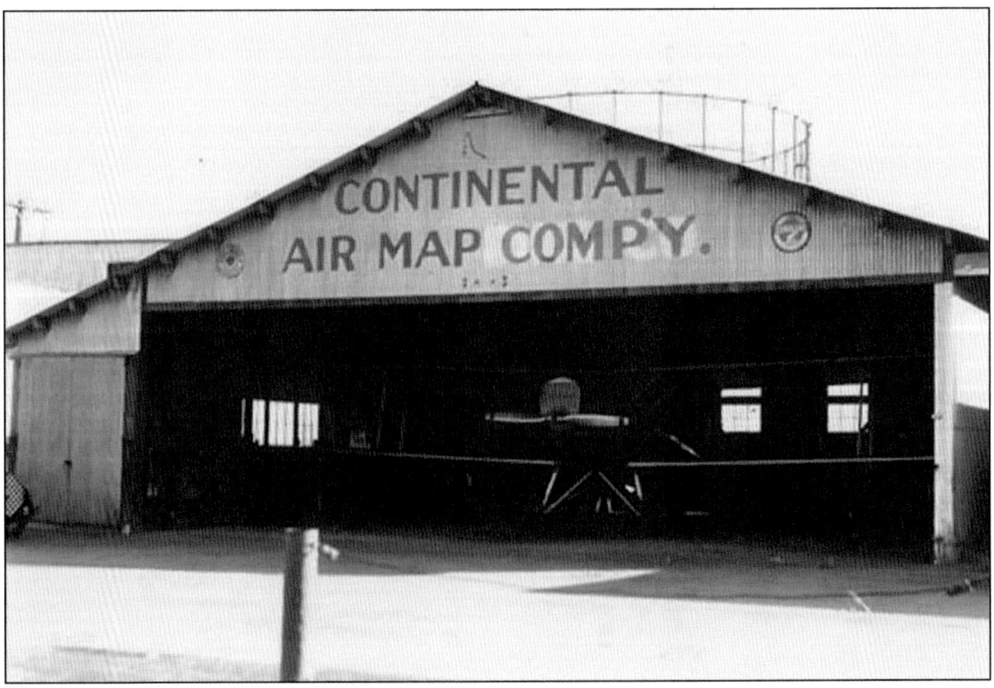

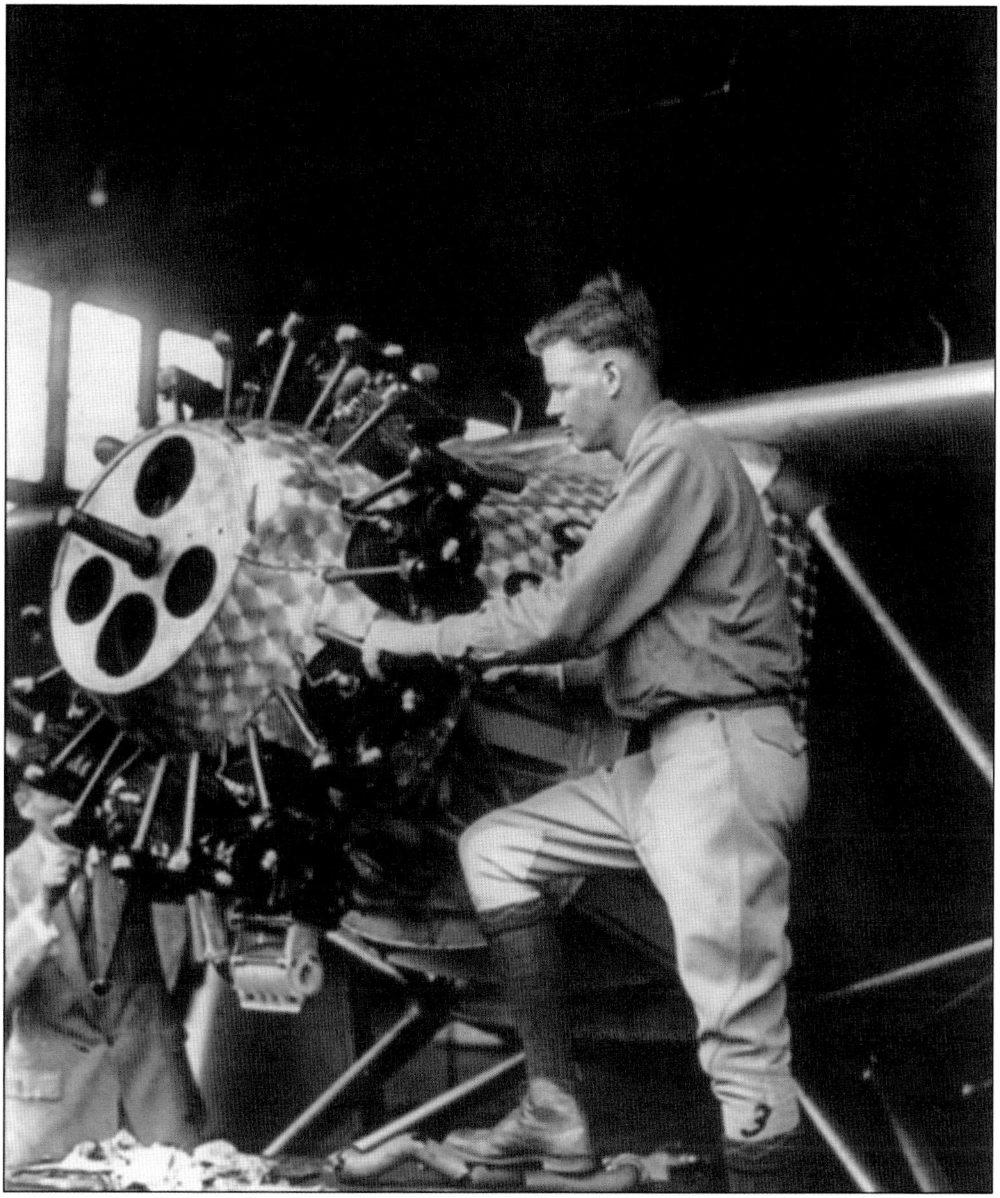

SPIRIT OF ST. LOUIS. T. (Tubal) Claude Ryan, a barnstormer who also trained at March Field, opened mail carrier service at Long Beach in 1926. Ryan's company used the Douglas "Cloudster" because of its design. Ryan Airlines was the first regularly scheduled passenger service between Los Angeles and San Diego. Ryan eventually sold his interest in his company but continued to serve as manager for Ryan Aeronautics, which produced the Ryan NYP that evolved from his M-1 design. The Ryan NYP was built by a team that included Douglas Corrigan, who worked for Ryan Airlines as a station manager. Investors purchased the plane at a cost of $10,580 for Charles A. Lindberg's planned trans-Atlantic flight between New York and Paris. Named *Spirit of St. Louis*, it was constructed as a "cabin plane" with a metal fuselage and a 220-horsepower, 9-cylinder, air-cooled Wright Aeronautical Whirlwind engine. Lindbergh (pictured with his plane), a former barnstormer and airmail pilot, would land at Long Beach on several occasions. So would Corrigan. (LOC.)

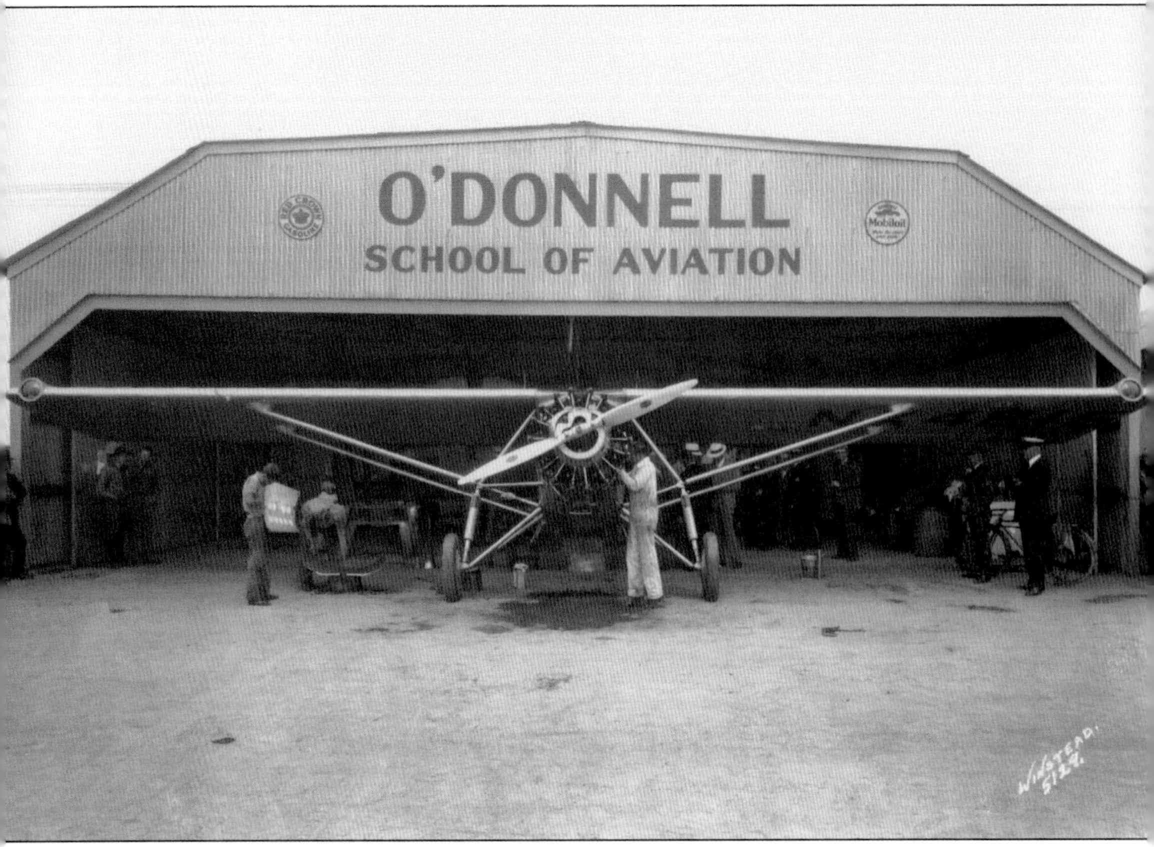

LINDBERG VISITS. In 1927, the city purchased an additional 125 acres for the airport and field in order to accommodate a longer runway. Inez B. Donavan, secretary of the Long Beach Chapter of the National Aeronautical Association, wrote about the airport: it was " well-marked with sock, circle and landing tee, and the 11-foot letters painted on the roof of one of the large hangars to identify the field are plainly visible from an altitude of 2,000 feet. The size of the field, the perfect drainage and the even air currents which prevail give Long Beach an airport that is unequaled for an all-year field, not only for air mail, passengers and freight, but also for factory sites. As it is water-bearing land and cannot be sold, lessees are assured of sites that will not be disposed of or sub-divided." Lindbergh's success increased national attention on aviation and flyers' interest in setting other distance records. While being feted in Long Beach after his solo flight, Lindbergh asked Lloyd O'Donnell to make minor repairs on his *Spirit of St. Louis*. O'Donnell ran a flight school and aviation repair shop and could not resist the temptation to fly Lindbergh's plane, charging several people $60 a person a ride.

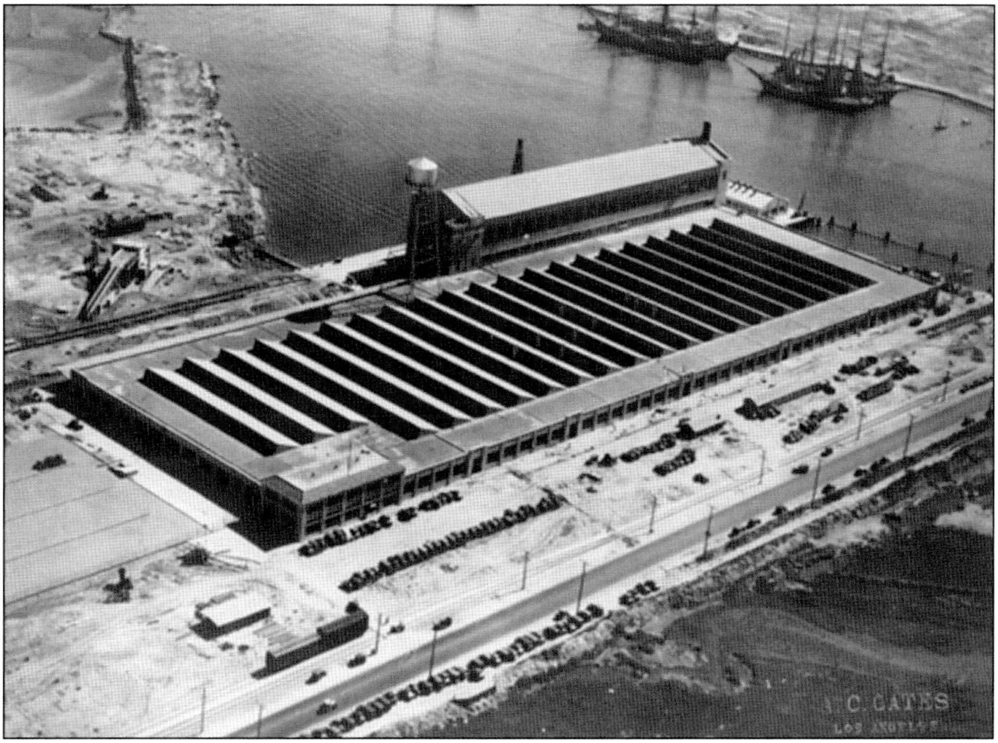

FORD COMES TO LONG BEACH. Henry Ford told city officials of his interest in building airplanes and his need for a larger local airport, giving Long Beach one more reason to develop a municipal airport. While Ford produced planes and plane motors elsewhere, the Long Beach plant only produced automobiles. The auto plant opened in April 1930 as the 35th assembly plant built by Ford. The Model A Ford was produced until 1931 when the plant switched to producing a V-8. The plant shut down in December 1932 and reopened in 1934. It was used by the U.S. Army Air Corps as a supply depot from 1934 until 1942. The plant resumed operations in 1945 and moved its operations to Pico Rivera in 1959. (Both, LOC.)

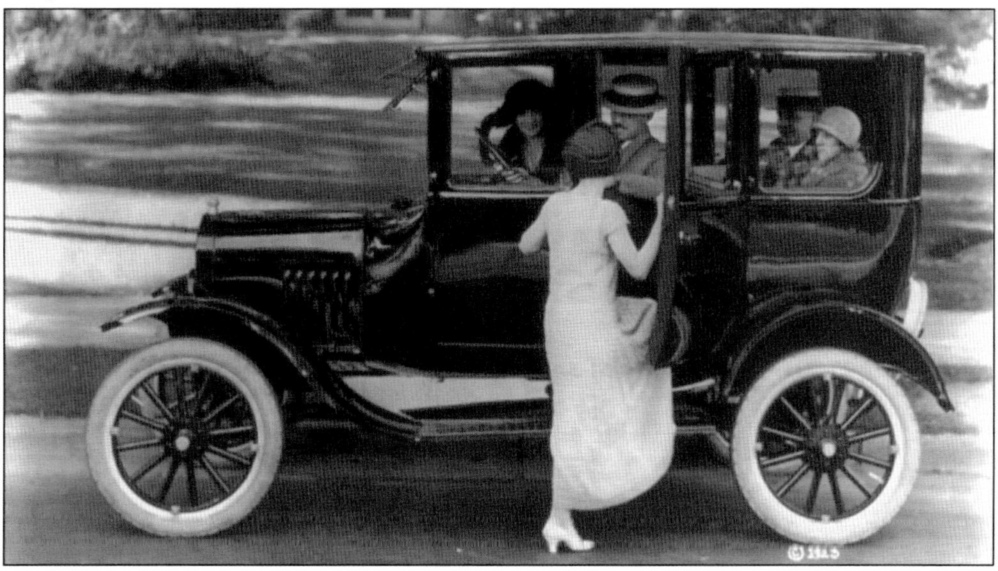

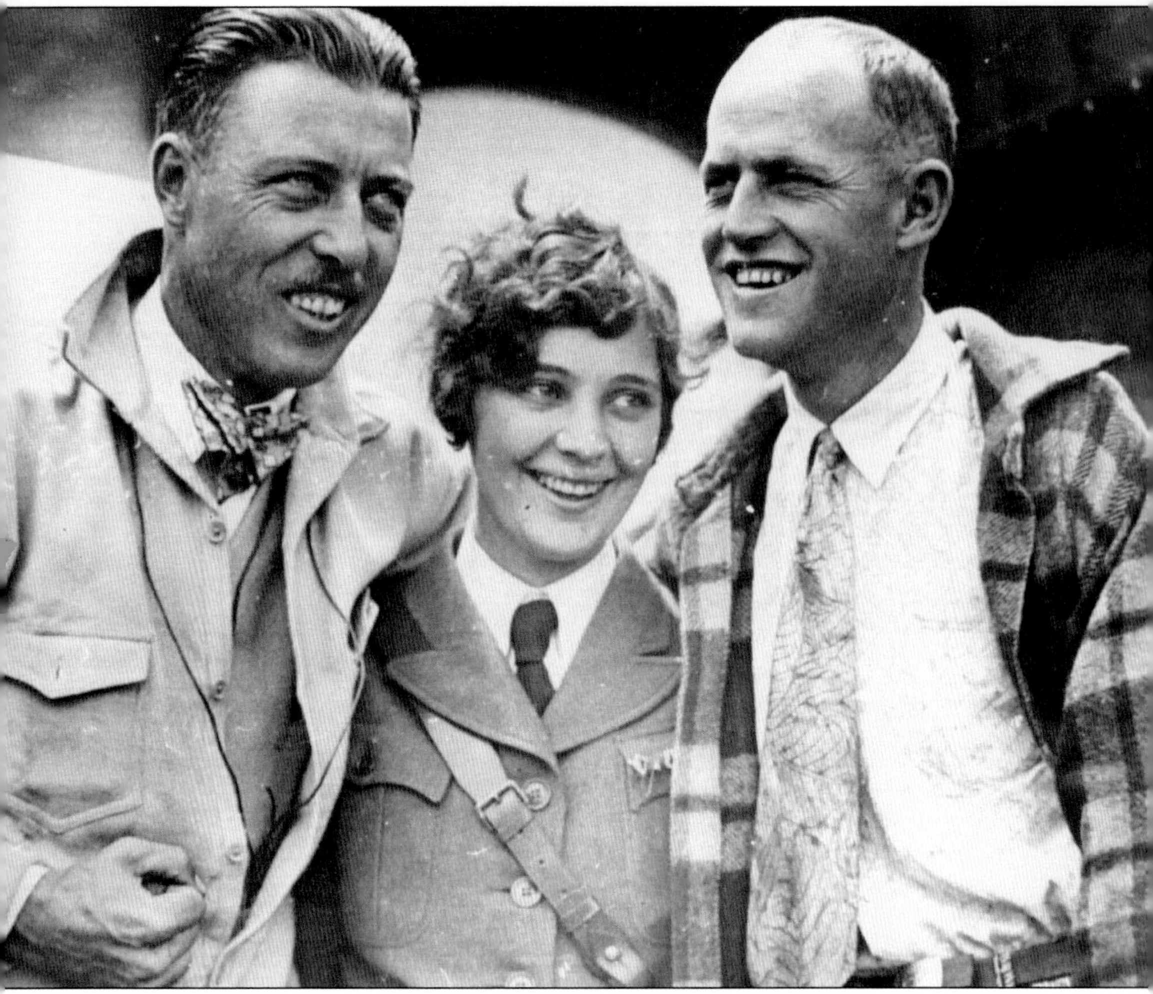

Auggy Pedlar. On August 16, 1927, more than 100,000 people watched as 15 planes took off from Oakland and headed toward Hawaii in the Dole Air Race. James Dole, heir to Dole Pineapple, put up a first prize of $25,000 and a second prize of $10,000. More than 15 pilots attempted to enter the race, but only eight planes took off, including the *Miss Doran* piloted by Auggy Pedlar (left), who had been a student of Earl Daugherty. The plane was navigated by Lt. Vilas Knope (right). The 24-year-old Pedlar flew the *Miss Doran*, named in honor of his passenger, Mildred Doran (center), a schoolteacher from Flint, Michigan. Doran asked to go along so she could claim being the "first woman to fly from the West Coast to Hawaii." Newspapers dubbed her "the flying school ma'am." Long Beach resident, attorney, and retired captain J. L. Giffin, his navigator Theodore Lundgren, and movie star Hoot Gibson flew *The Pride of Los Angeles* (an F-10 triplane built by the Long Beach International Aircraft Corporation) but crashed offshore Oakland. They survived. (HAPS.)

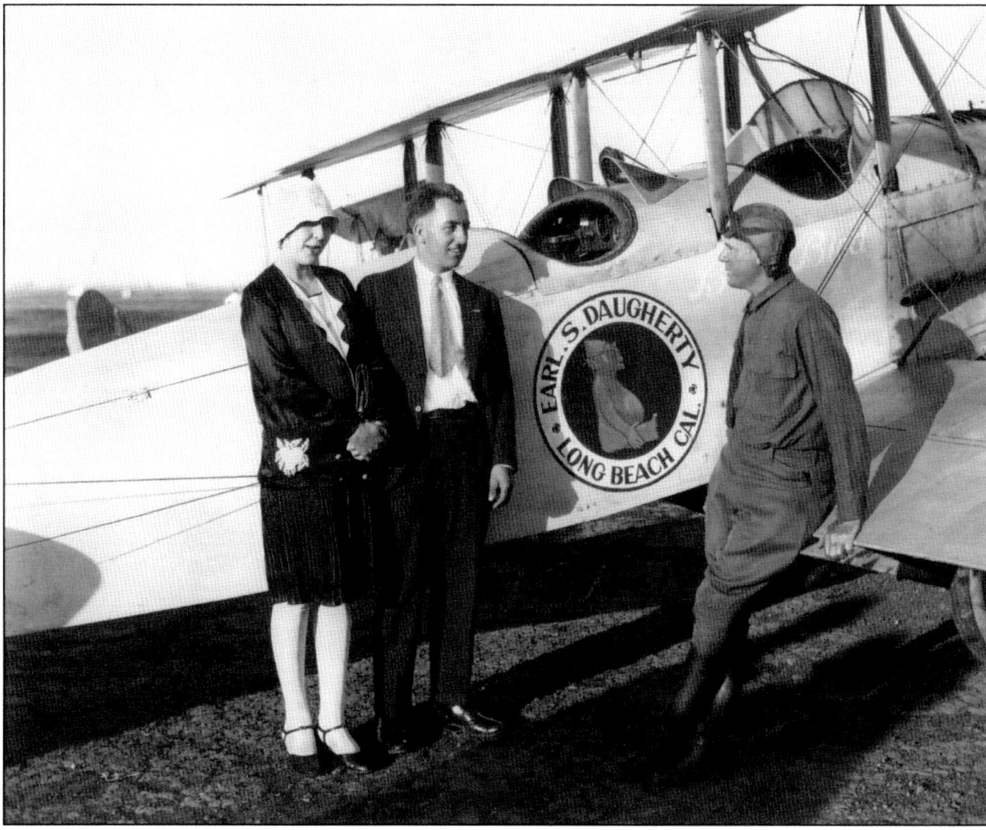

Billiken Power. Earl Daugherty was first given a "Billiken" in 1912 from a Virginia Hotel guest, a Mrs. Stewart. He carried one in his plane, etched one on his watchband and watch crystal, and painted one on the outside of his planes. The "Billiken" was a made-up character by art teacher Florence Pretz of Kansas City, Missouri, and marketed as a good-luck charm as the "god of things as they should be." The "Billiken" is actually named after the newly elected Pres. William Howard Taft and remained popular in the 1900s.

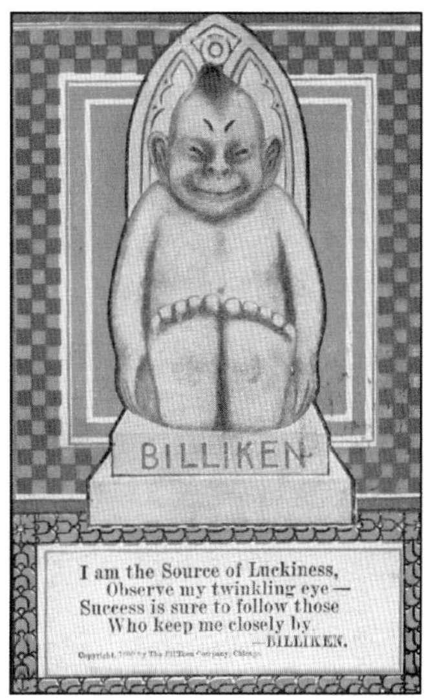

Bad Luck for Pedlar. The night before the Dole Race, Pedlar and Doran dined with Earl and Kay Daugherty in Long Beach. Daugherty gave Pedlar a gold Billiken to carry on his flight to Hawaii. The *Oakland Tribune* reported, "Pedlar thinks the famous 'Billiken' smile will give him courage long after his own grin wears off in the desperate struggle to beat competitors and forces of nature on the 2400 mile air jaunt to hula land." Daugherty's Billiken didn't bring Pedlar good luck. The *Miss Doran* took off and experienced engine trouble. Pedlar returned to the field, fixed the problem, and took off; the *Miss Doran* was never seen again. A total of 11 people lost their lives in the Dole Air Race. (AC.)

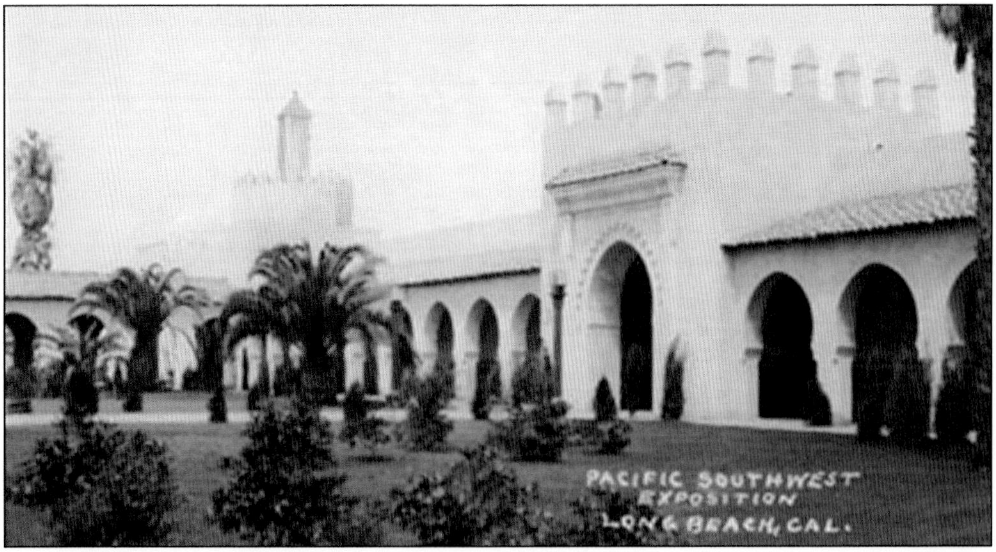

THE 1928 EXPOSITION. July 27 through August 13, 1928, Long Beach hosted the largest exposition in California since the 1915 Panama Pacific Exposition at San Francisco and San Diego. Thousands came to view the Pacific Southwest Exposition (above and below). Ten halls on 65 acres of Long Beach waterfront were filled with exhibits from cities throughout the Pacific Southwest as well as 33 countries, such as Belgium, Mexico, Germany, Guatemala, Japan, and Sweden. Fifty-five buildings constructed in a "Moorish motif" provided exhibition space. Airplane and dirigible demonstrations were featured. Ten naval destroyers and eight battleships were anchored off the coast. The exposition was visited by Pres. Herbert Hoover, Charles Lindbergh, and the *Our Gang* actors from Fox Film. (Both, LBPL.)

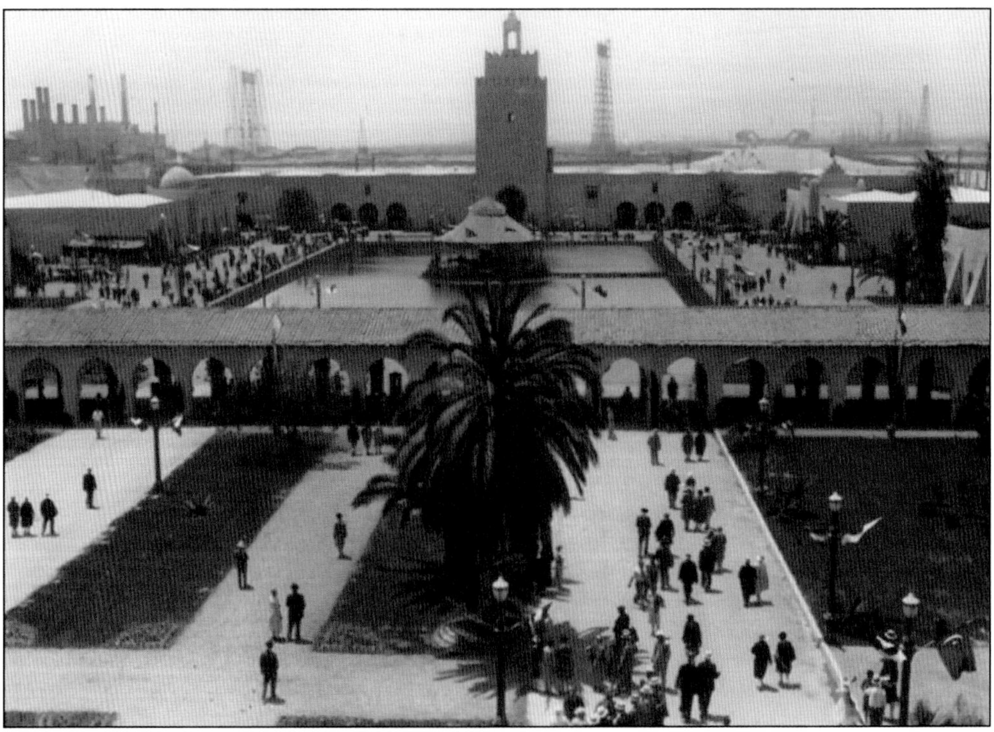

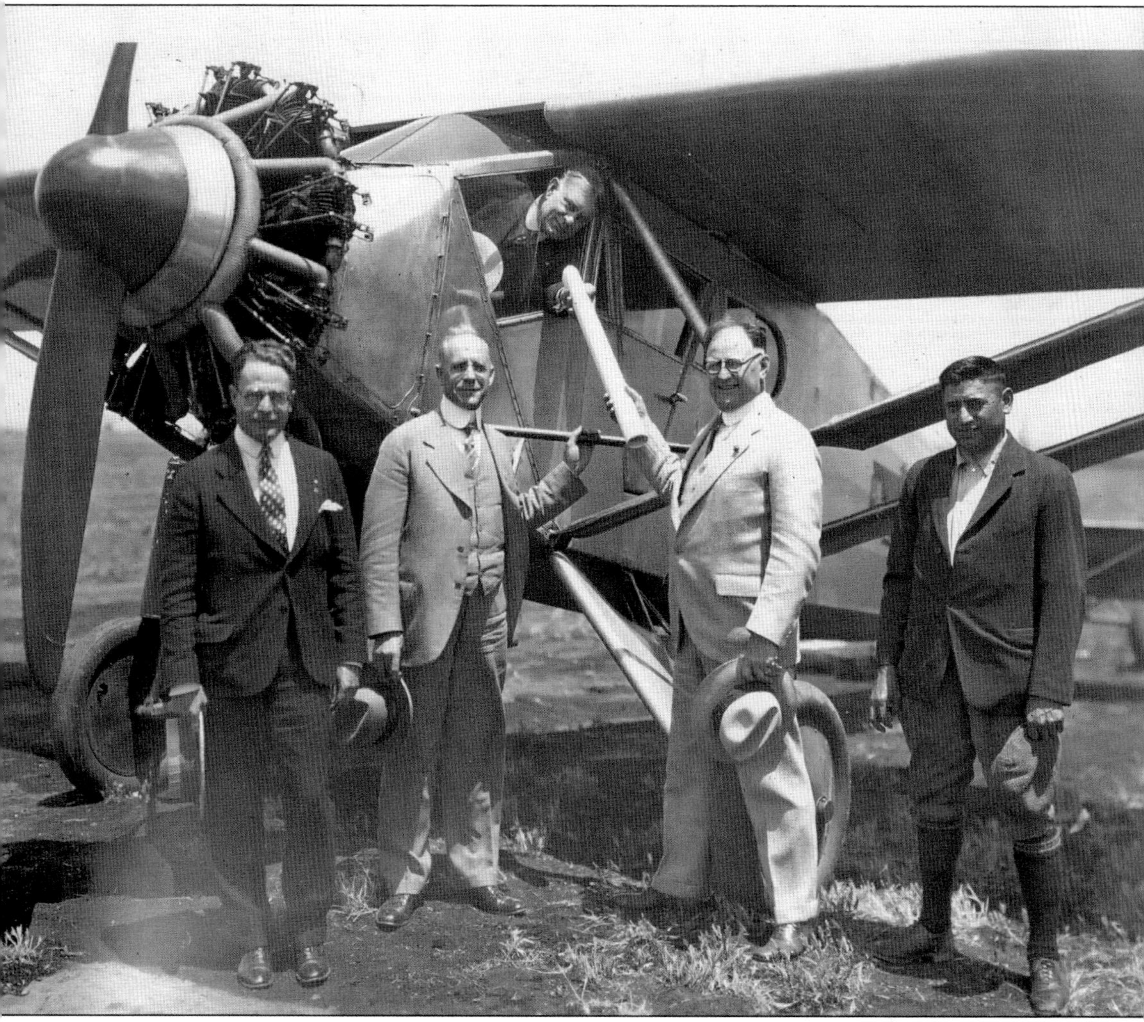

SELLING LONG BEACH. Hoping to promote the Pacific Southwest Exhibition, the Long Beach City Council and other officials accepted John Montijo's invitation to join him and business partner C. B. "Boone" Bellows in their "Belmont Cabin Monoplane C/N M-1." The delegation included, from left to right, Harry C. Morgan, councilman; J. J. Barton, city manager; H. S. Callahan (in cabin); city clerk J. Oliver Brison; and John G. Montijo, pilot. They flew to Bakersfield, to San Diego and Santa Barbara, to San Bernardino and El Centro, and ended in Elsinore. They dropped pamphlets good for free admission to the exposition. San Diego officials entertained the group at a banquet luncheon and an automobile trip around the city. Betty Grey, "copilot" with Montijo, dressed in a colorful Spanish costume symbolizing the Spirit of Old Spain, which was the theme of the exposition. After the tour, Brison commented to the press, "On the whole, the tour accomplished its purpose 100 percent."

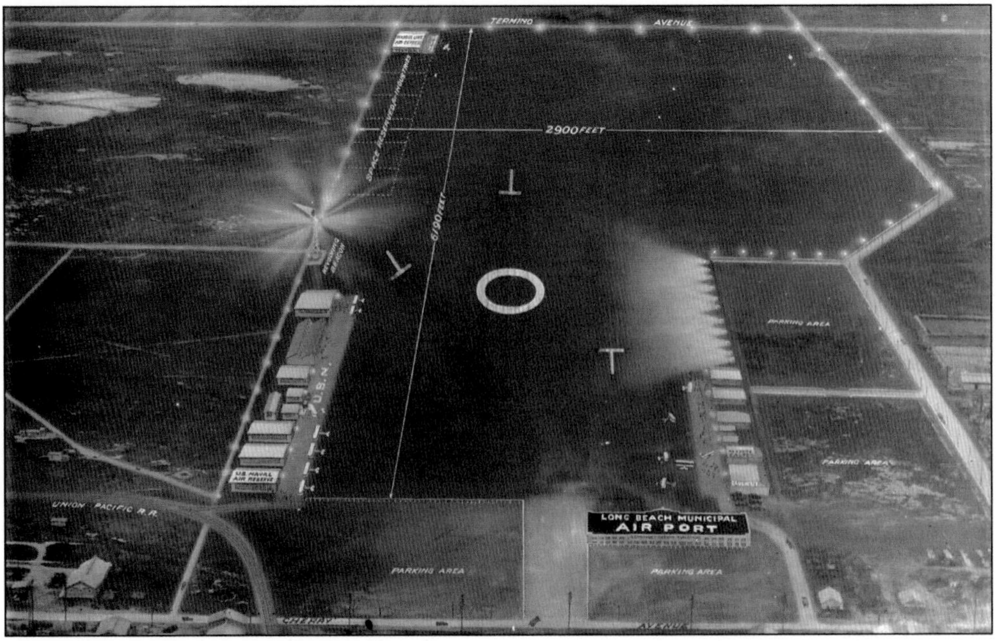

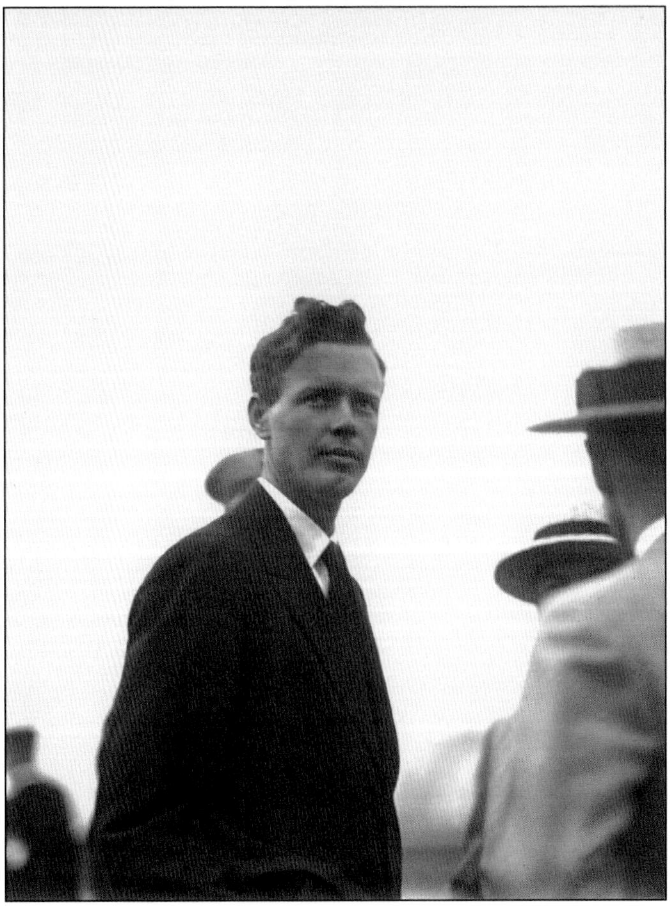

Lights on for Lindy. Eventually lights would be installed surrounding the airport, as shown in this drawing of the field. In 1928, Lindbergh made an emergency landing after midnight with his passenger Col. Henry Breckenridge. A Long Beach resident heard a plane flying over the field and made a telephone call to the airport, which resulted in a light being turned on, allowing Lindbergh to see the field and land. A grateful Lindbergh later returned and publicly lauded the city for its state-of-the-art features such as paved runways and night lighting. He dined with the airport field superintendent and stayed the night at the Breakers Hotel on Ocean Boulevard. Newspaper stories indicate Lindbergh returned several times to Long Beach and stayed at the Breakers.

DAUGHERTY AND WIFE BUY A PLANE. In late 1928, Earl Daugherty had another close call. The Curtiss Robin plane he was copiloting with D. R. Robertson exploded in midair, sending the plane to the ground on a farm in Carthage, Indiana, injuring Robertson. Daugherty was unhurt. As a result, Daugherty could not finish the September 1928 transcontinental air race between New York and Los Angeles. A few months later, Daugherty and his wife, Kay, purchased another plane, a Whirlwind Laird, in Chicago at a cost of $11,000.

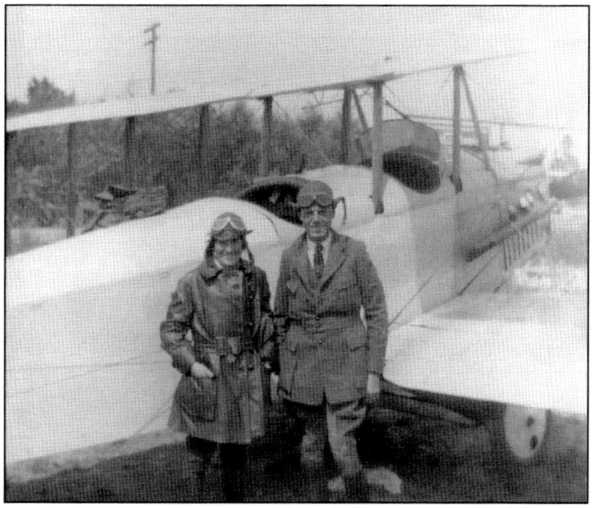

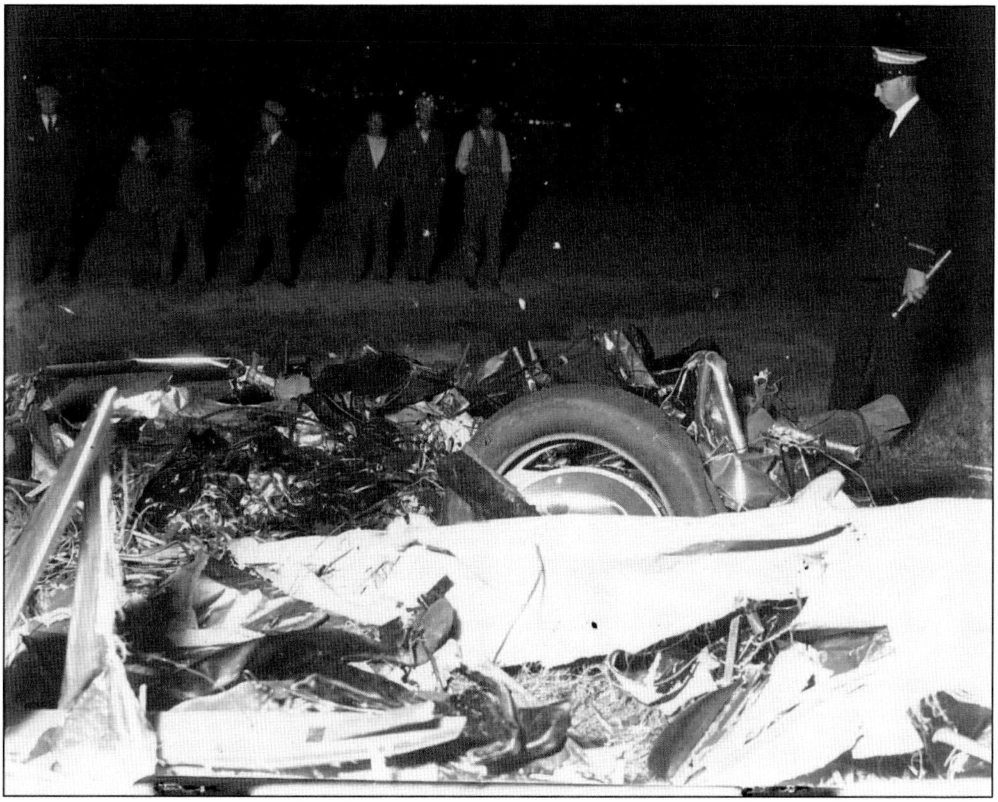

DAUGHERTY DIES. December 8, 1928, Earl Daugherty flew his new plane for the last time. Within moments of reaching 1,500 feet, the wings of the plane collapsed, sending it into a nose dive to the ground one mile east of Cherry Avenue and a quarter mile north of Wardlow Avenue (pictured). Daugherty, Willis Montfort (city editor of the *Press Telegram*), and Elmer Starr (head of the Pacific Engraving Company) perished as hundreds of spectators looked on in horror. Montfort and Starr frequently took rides with Daugherty over the city. Because Daugherty had the plane serviced earlier in the week, his Billiken was removed from the plane and his watch bearing a Billiken was in the shop. His wife, Kay, placed a Billiken in his burial crypt.

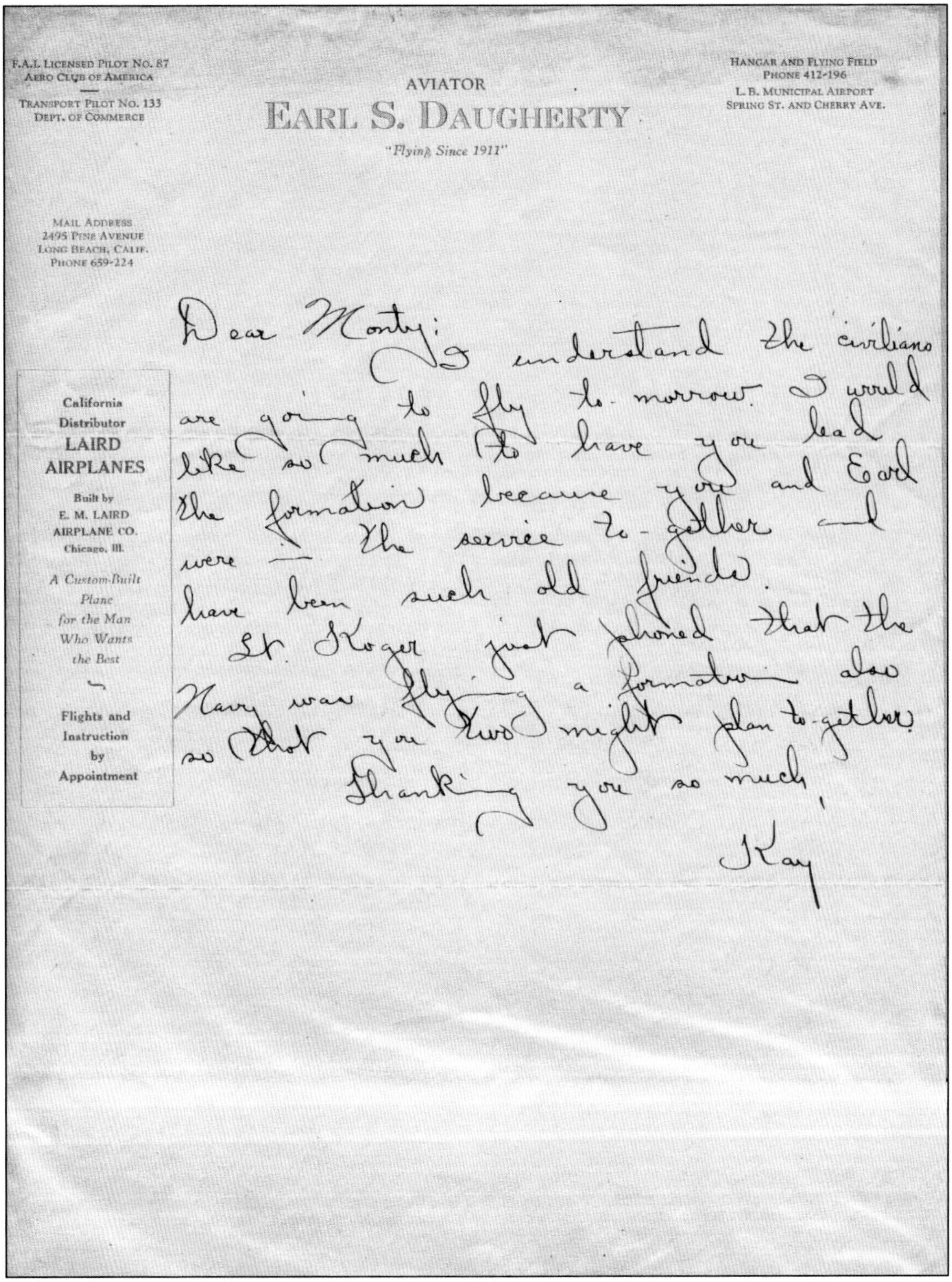

F.A.I. Licensed Pilot No. 87
Aero Club of America

Transport Pilot No. 133
Dept. of Commerce

AVIATOR
EARL S. DAUGHERTY
"Flying Since 1911"

Hangar and Flying Field
Phone 412-196
L. B. Municipal Airport
Spring St. and Cherry Ave.

Mail Address
2495 Pine Avenue
Long Beach, Calif.
Phone 659-224

California
Distributor
LAIRD
AIRPLANES
Built by
E. M. LAIRD
AIRPLANE CO.
Chicago, Ill.
A Custom-Built
Plane
for the Man
Who Wants
the Best

Flights and
Instruction
by
Appointment

Dear Monty: I understand the civilians are going to fly to-morrow. I would like so much to have you lead the formation because you and Earl were in the service to-gether and have been such old friends.

Lt. Roger just phoned that the Navy was flying a formation also so that you two might plan to-gether.

Thanking you so much,

Kay

Daugherty Is Buried. Daugherty was buried at the Angeles Abbey Mausoleum, just a few miles from both Long Beach and Dominquez airfields. His widow, Kay, wrote John Montijo and asked that he head the formation of planes to accompany the funeral procession. Daugherty's funeral was attended by numerous flyers. In honor of Daugherty's contributions, the city council later renamed the airport the Long Beach Municipal Airport–Daugherty Field. In 1930, the Daugherty Park Golf Course at American Avenue and Willow Street was dedicated. The city council accepted a gift of a portion of "the old Daugherty Field" as a park and a playground.

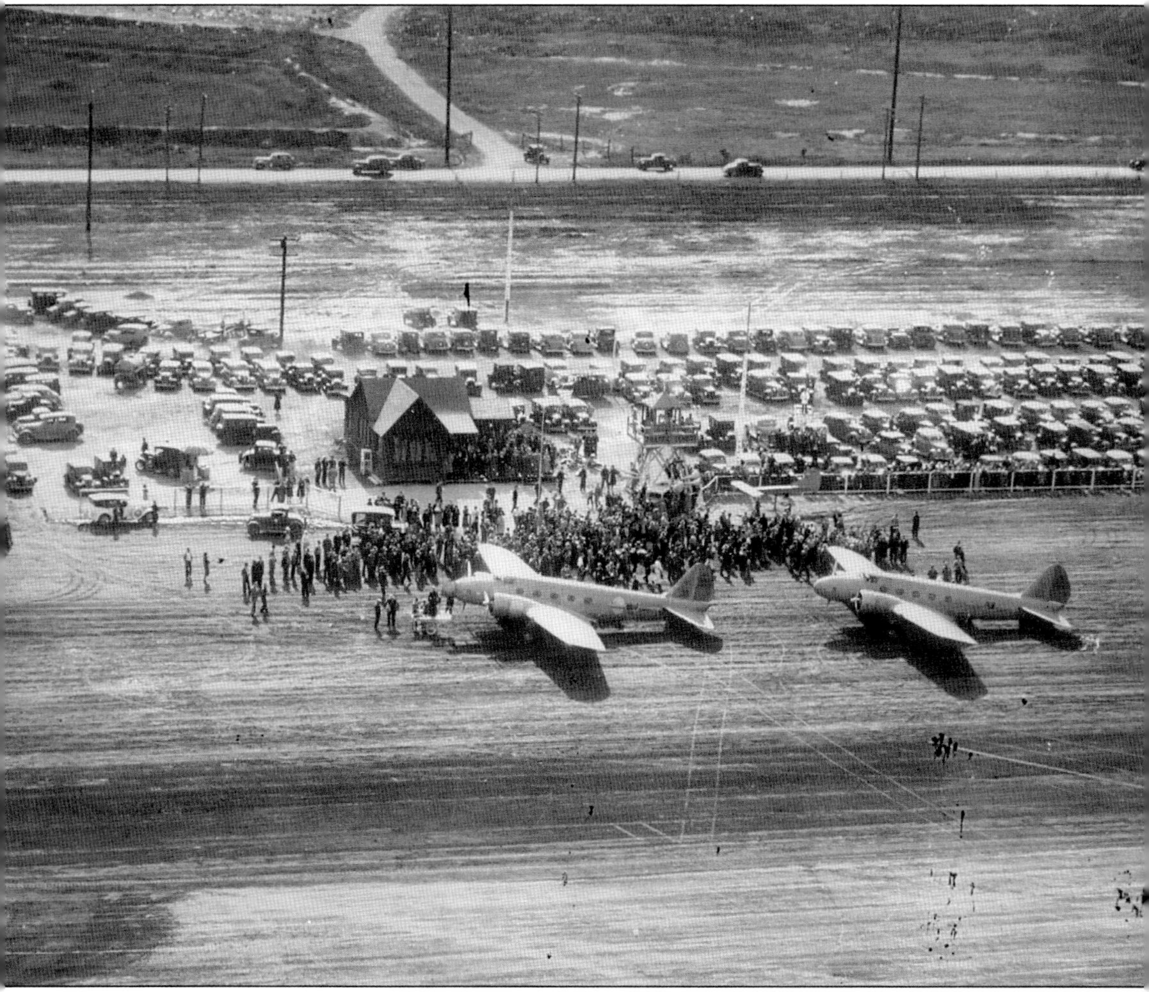

Commercial Aviation. Thanks to Earl Daugherty's persistence, Long Beach was one of the first cities to have a municipal airport, preceding Los Angeles by more than three years. Long Beach fiercely competed with Mines Field in Los Angeles (which would become LAX) during the 1930s and 1940s to establish this city as the air terminal of the Pacific Coast. Commercial airlines made investments in permanent installations at Glendale and Burbank and preferred the Los Angeles airfield instead of Long Beach because of the number of unscheduled small, private, and military aircraft that took off from the field. The city council and chamber of commerce focused their energies on making the municipal airport a site for commercial aviation. Airport records indicate that Western Air Lines (pictured) first carried passengers, mail, and cargo from the Long Beach Airport in September 1929.

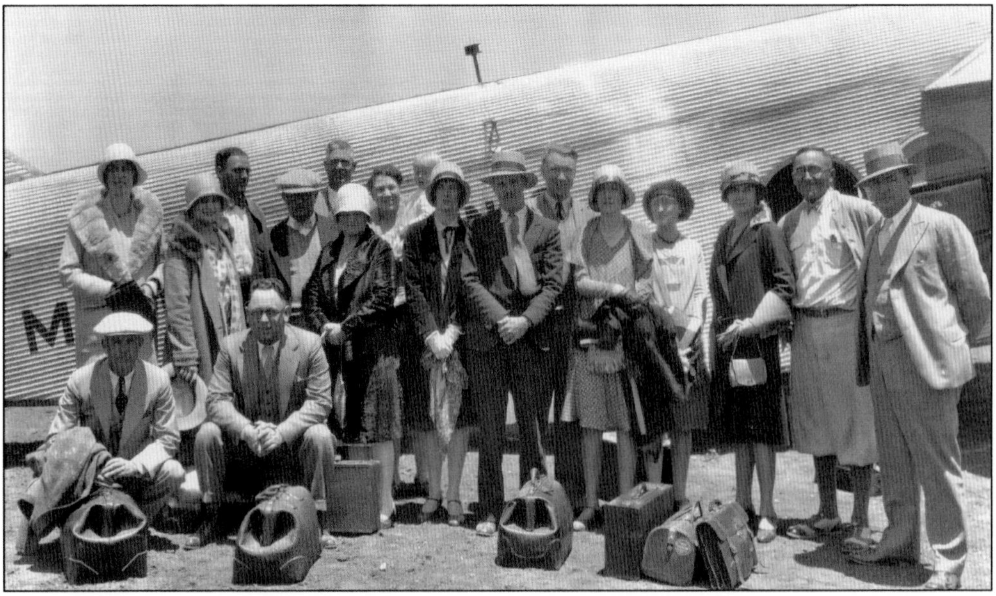

Long Beach to Agua Caliente. Aviation exploded in Southern California with airfields to the north, south, east, and west of Long Beach. Flying schools and aircraft manufacturers were established, such as the Long Beach International Aircraft Corporation at Spring Street and Pacific Avenue, which produced the F-17 and the F-10. Maddux Airlines flew out of Long Beach to Agua Caliente, Mexico, for $37.50 round-trip. By 1931, Long Beach installed boundary lights on the north and east sides of the airport. Gilpin Airlines offered service to Los Angeles and San Diego. In 1937, radio and weather reporting facilities were installed so that pilots could fly into Long Beach on radio beam.

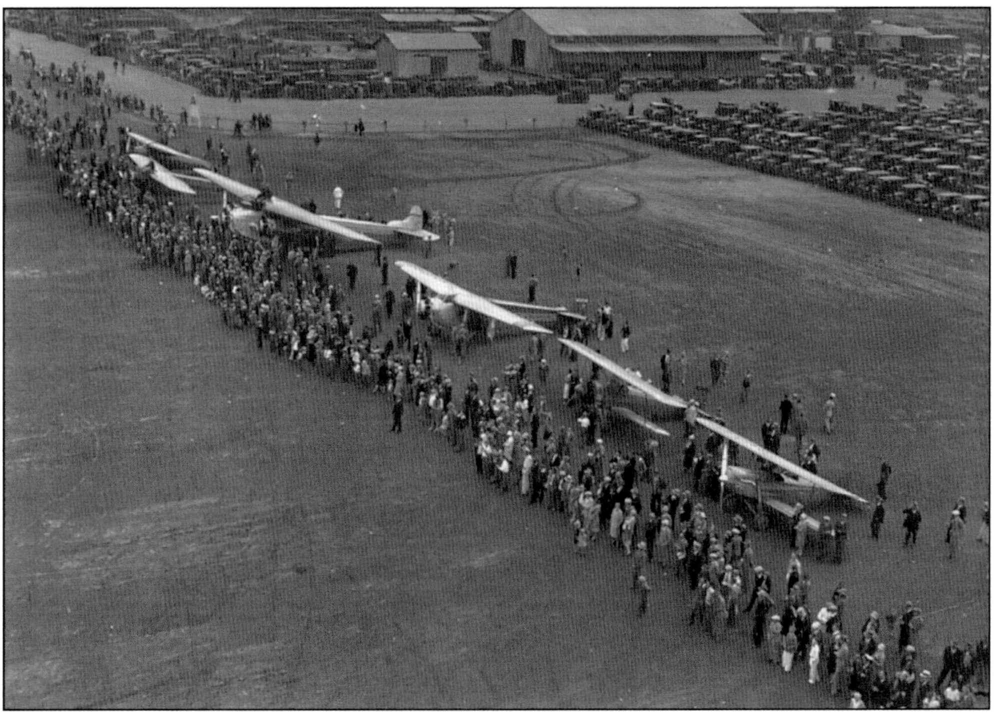

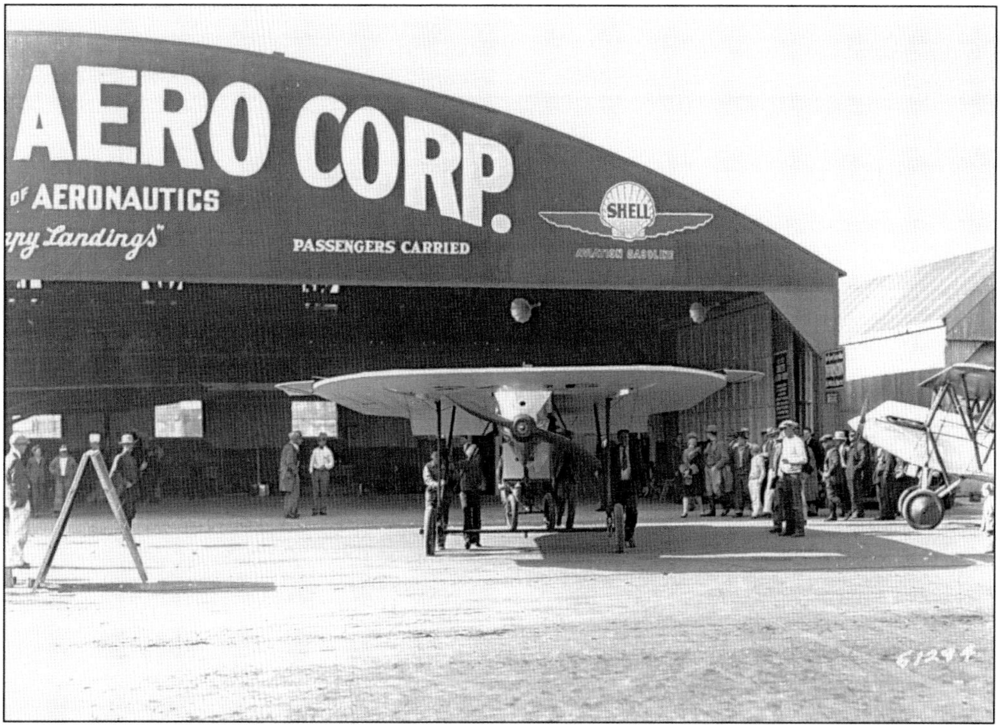

McLeary Experimental and Auto-Giro. After airport expansion, Squire Du Ree told the press he would recommend to the city manager and council leasing of the aviation field space to private aviators, "eighteen of whom have signified their willingness to pay a rental for facilities at the field." Hangars were built and leased as shown above. The field attracted airplanes of all types, including the experimental McLeary Airplane (above) and the Pitcairn "Auto-Giro" (shown below with John Montijo at the Grand Central Air Terminal in Glendale), which Amelia Earhart demonstrated in Long Beach in 1931. City officials wanted to make Long Beach an airport "second only in size to the field at Oakland."

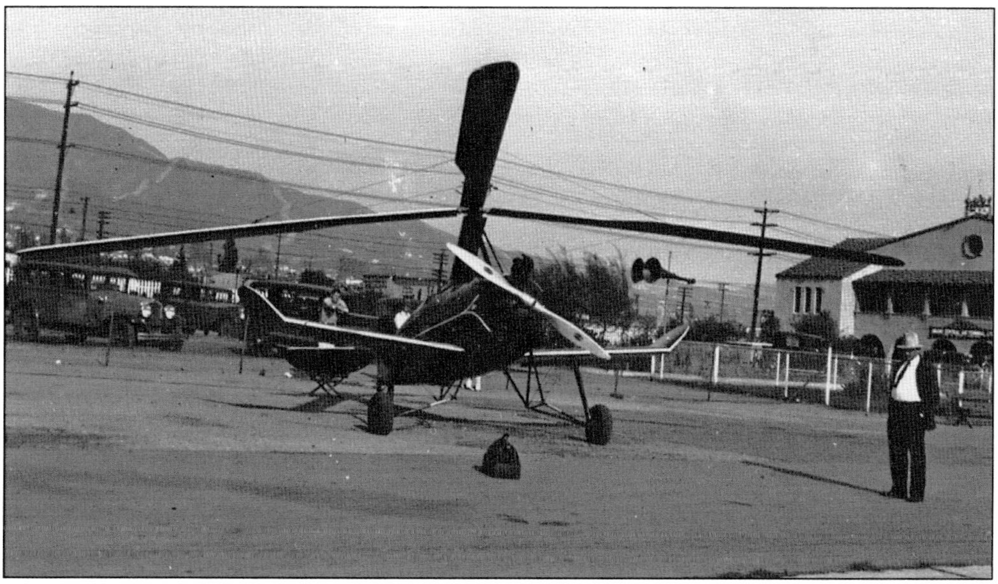

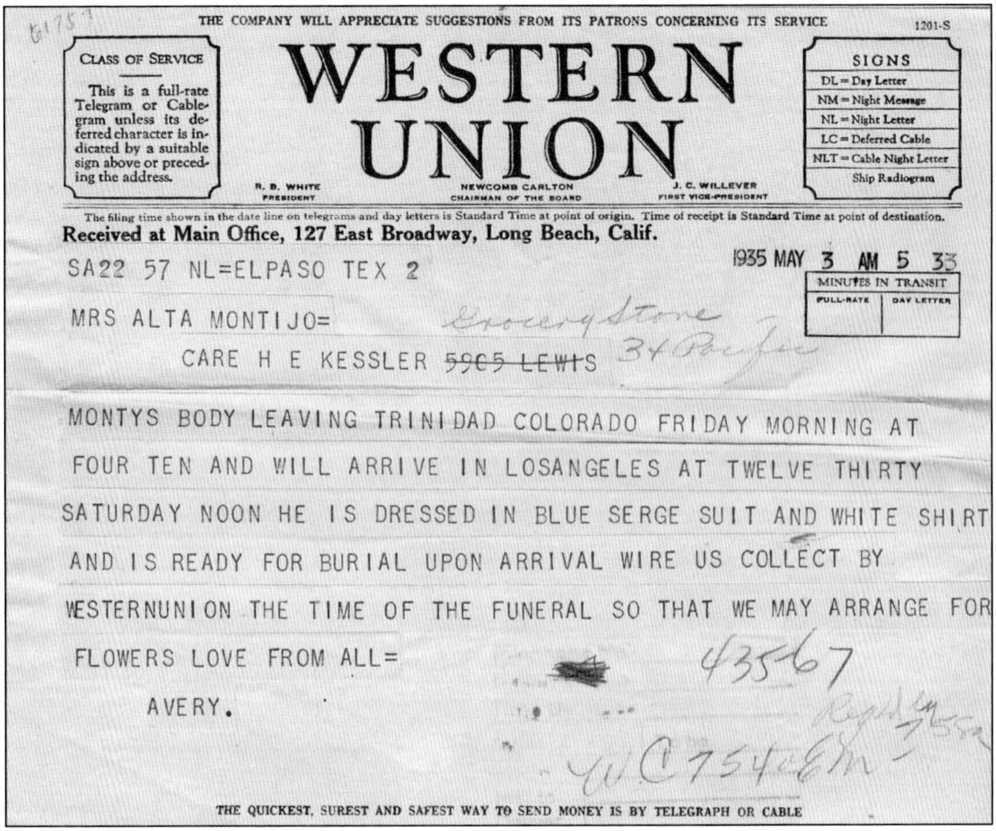

THE COMPANY WILL APPRECIATE SUGGESTIONS FROM ITS PATRONS CONCERNING ITS SERVICE 1201-S

WESTERN UNION

CLASS OF SERVICE — This is a full-rate Telegram or Cablegram unless its deferred character is indicated by a suitable sign above or preceding the address.

R. B. WHITE, PRESIDENT — NEWCOMB CARLTON, CHAIRMAN OF THE BOARD — J. C. WILLEVER, FIRST VICE-PRESIDENT

SIGNS
DL = Day Letter
NM = Night Message
NL = Night Letter
LC = Deferred Cable
NLT = Cable Night Letter
Ship Radiogram

The filing time shown in the date line on telegrams and day letters is Standard Time at point of origin. Time of receipt is Standard Time at point of destination.

Received at Main Office, 127 East Broadway, Long Beach, Calif.

SA22 57 NL=ELPASO TEX 2 1935 MAY 3 AM 5 33

MINUTES IN TRANSIT — FULL-RATE | DAY LETTER

MRS ALTA MONTIJO=
CARE H E KESSLER ~~5905 LEWI~~S

MONTYS BODY LEAVING TRINIDAD COLORADO FRIDAY MORNING AT FOUR TEN AND WILL ARRIVE IN LOSANGELES AT TWELVE THIRTY SATURDAY NOON HE IS DRESSED IN BLUE SERGE SUIT AND WHITE SHIRT AND IS READY FOR BURIAL UPON ARRIVAL WIRE US COLLECT BY WESTERNUNION THE TIME OF THE FUNERAL SO THAT WE MAY ARRANGE FOR FLOWERS LOVE FROM ALL=

AVERY.

THE QUICKEST, SUREST AND SAFEST WAY TO SEND MONEY IS BY TELEGRAPH OR CABLE

MONTIJO DIES. John Montijo died May 1, 1935, while delivering mail between Pueblo, Colorado, and El Paso, Texas. Trying to swoop low over a school yard near Wallenburg, Colorado, to drop a note to the students below who had been corresponding with him, his plane dipped and a wing hit the ground, plummeting him into the ground. His note asked the students to let him know if any Easter eggs he had dropped on another trip had broken. A teacher and 12 students witnessed the crash. Montijo's body was shipped back to Long Beach via train by Varney Airlines. Montijo's funeral was held in Long Beach, and 20 planes followed the hearse to Angeles Abbey Mausoleum in Compton, California, where his friend Earl Daugherty (pictured left of Montijo) had been buried. Montijo's sons Jimmy and Jack later served as pilots during World War II.

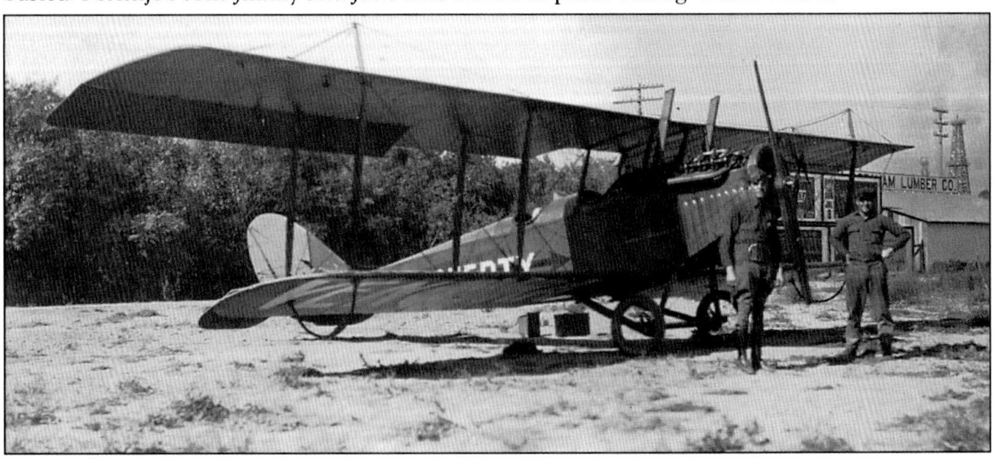

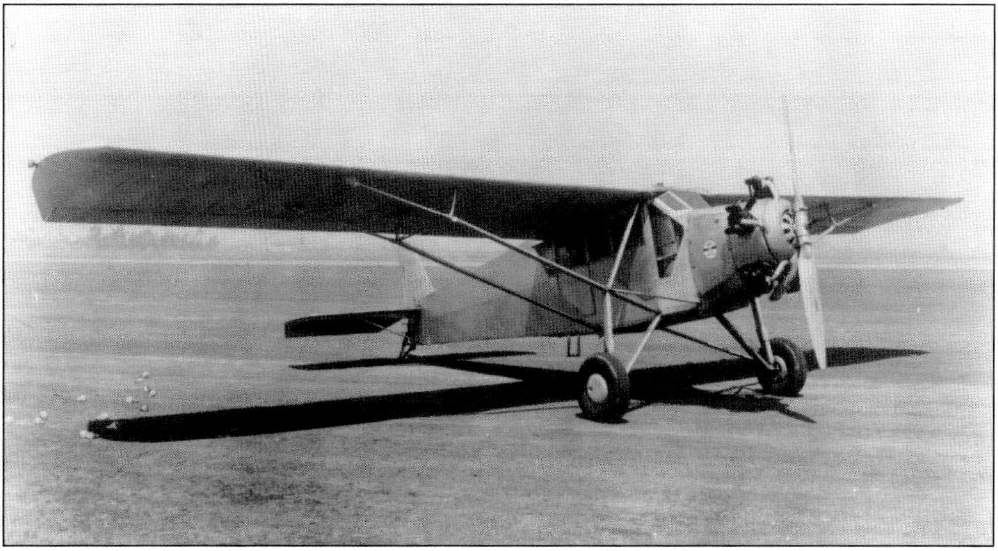

Wrong Way Corrigan. Douglas Corrigan, station manager for Ryan Airlines, worked on the assembly of Charles Lindbergh's *Spirit of St. Louis*. Wanting to duplicate Lindbergh's success, Corrigan dumbfounded the aviation world when he took off from Long Beach in 1938 and flew nonstop to New York in a secondhand plane called the "$900 crate." Instead of turning around and flying back to Long Beach, Corrigan headed for Ireland. Upon his arrival, he feigned not having a compass nor map and contended he had flown 28 hours "the wrong way," earning him the nickname "Wrong Way Corrigan." A cheering crowd greeted him when he returned to Long Beach. The city erected a plaque in his honor, which still hangs in the Municipal Airport terminal building. (Above, LOC.)

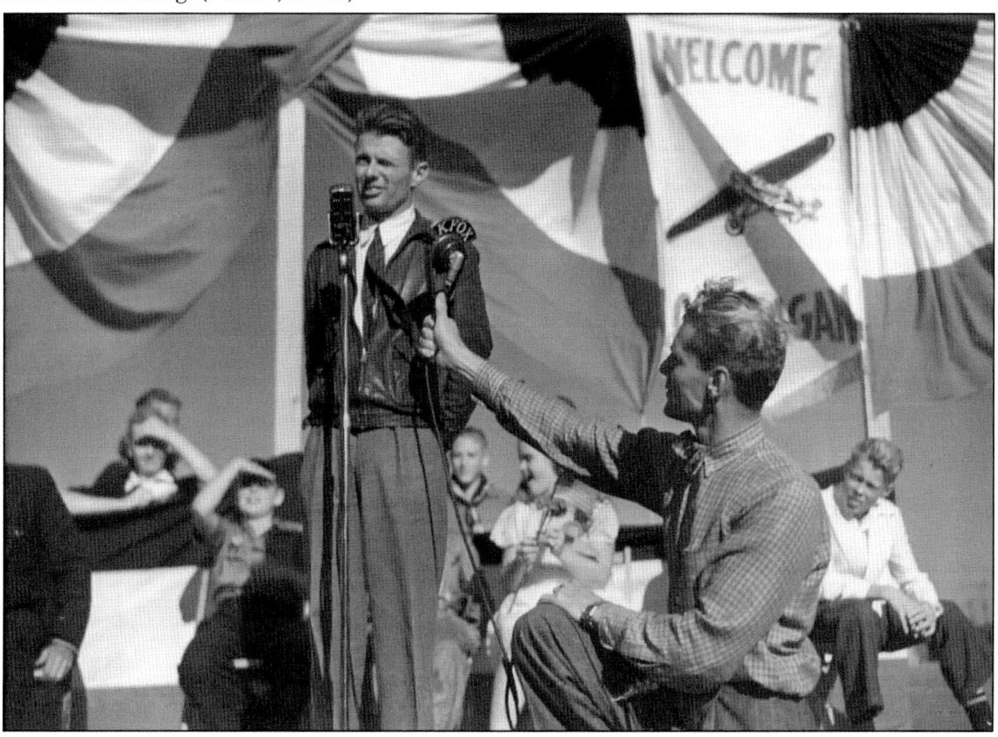

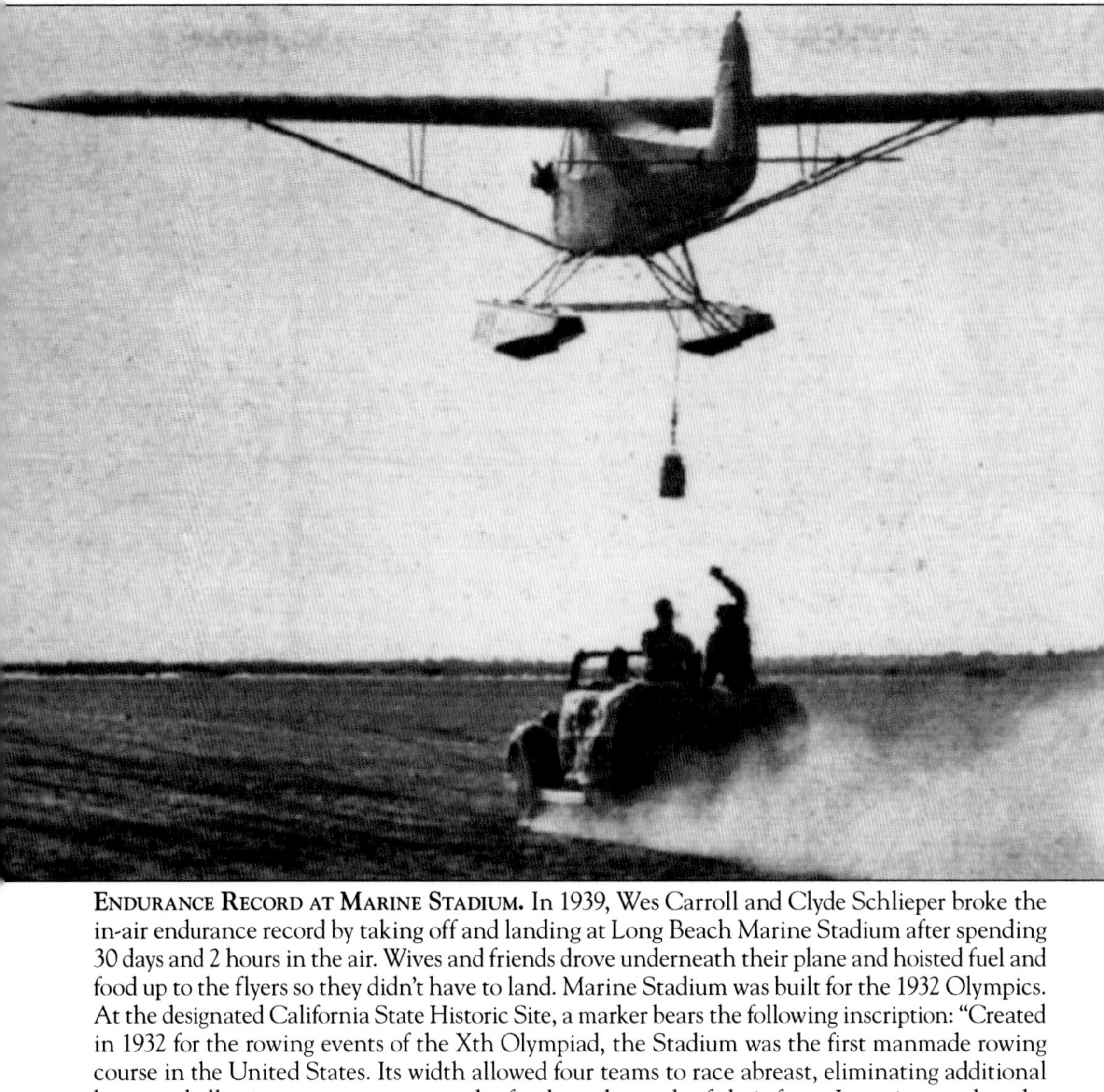

Endurance Record at Marine Stadium. In 1939, Wes Carroll and Clyde Schlieper broke the in-air endurance record by taking off and landing at Long Beach Marine Stadium after spending 30 days and 2 hours in the air. Wives and friends drove underneath their plane and hoisted fuel and food up to the flyers so they didn't have to land. Marine Stadium was built for the 1932 Olympics. At the designated California State Historic Site, a marker bears the following inscription: "Created in 1932 for the rowing events of the Xth Olympiad, the Stadium was the first manmade rowing course in the United States. Its width allowed four teams to race abreast, eliminating additional heats and allowing oarsmen to enter the finals at the peak of their form. Later it served as the venue for the 1968 and 1976 United States men's Olympic rowing trials and the 1984 United States women's Olympic rowing trials. The site remains an important training and competitive center for rowers, including our National and Olympic teams." (www.flightglobal.com.)

Seven

The Military and the Long Beach Municipal Airport

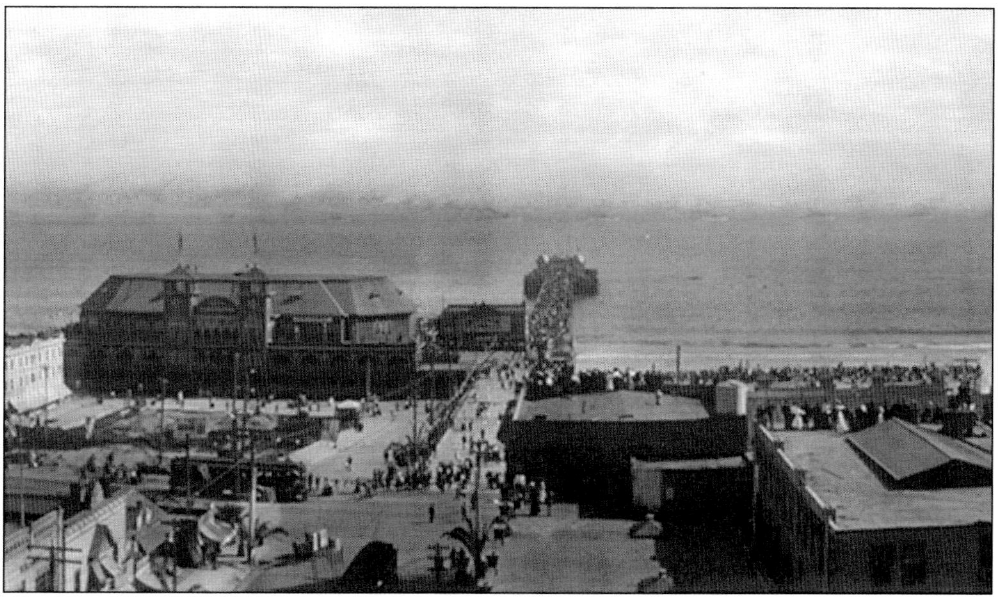

Naval Interest in Long Beach. December 16, 1907, Pres. Theodore Roosevelt sent 17 battleships of the 2nd Division fleet painted white—"The Great White Fleet"—around the world. The fleet stopped in Long Beach in April 1908, and 50,000 turned out on the beach to greet the fleet with a hosted barbecue. (LOC.)

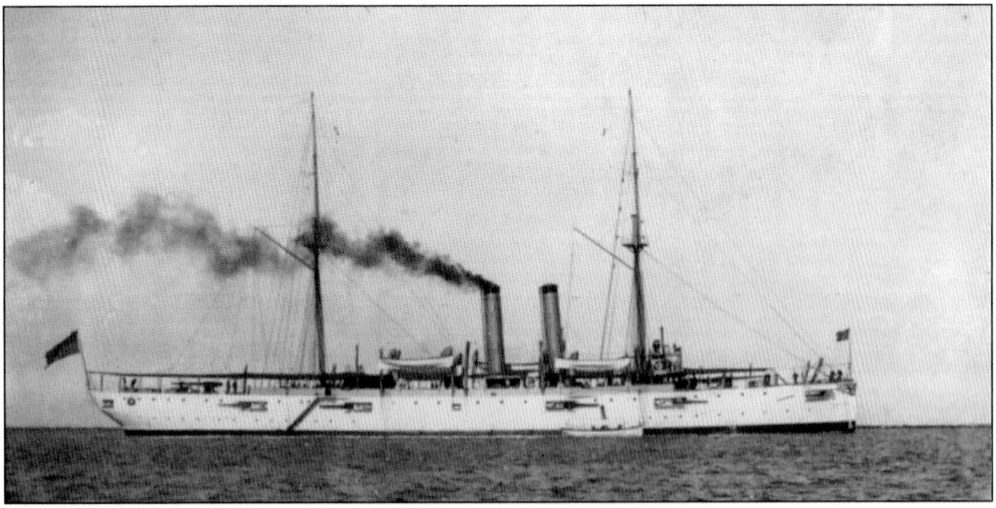

Navy Continues to Call. As early as 1907, city officials traveled to Washington, D.C., requesting a naval base in Long Beach because of "her harbor advantages," offering to "donate the site if necessary." The navy frequently anchored off Long Beach in the early 1900s, including on July 3, 1915, when the gunboat USS *Chattanooga* sent by naval secretary Josephus Daniels fired two 21-gun salutes in honor of Long Beach's dedication of a Lincoln Grand Army of Republic Monument in Pacific Park (below) and regrets on behalf of Pres. Woodrow Wilson. It was Daniels who ordered sailors to drink coffee instead of alcohol, thus creating the expression "cup of Joe." In March 1921, the "eighth spotting squadron" of the Pacific Fleet, comprised of 12 DeHavilland airplanes, arrived at the Long Beach harbor on a barge and was then taken to "Camp Daugherty Aviation Field" to participate in maneuvers and target practice off the coast of Long Beach. Tents were erected at the airfield for the pilots and the mechanics. Long Beach would later expend millions of dollars to construct naval landings to accommodate sailors coming ashore. (Above, LOC; below, AC.)

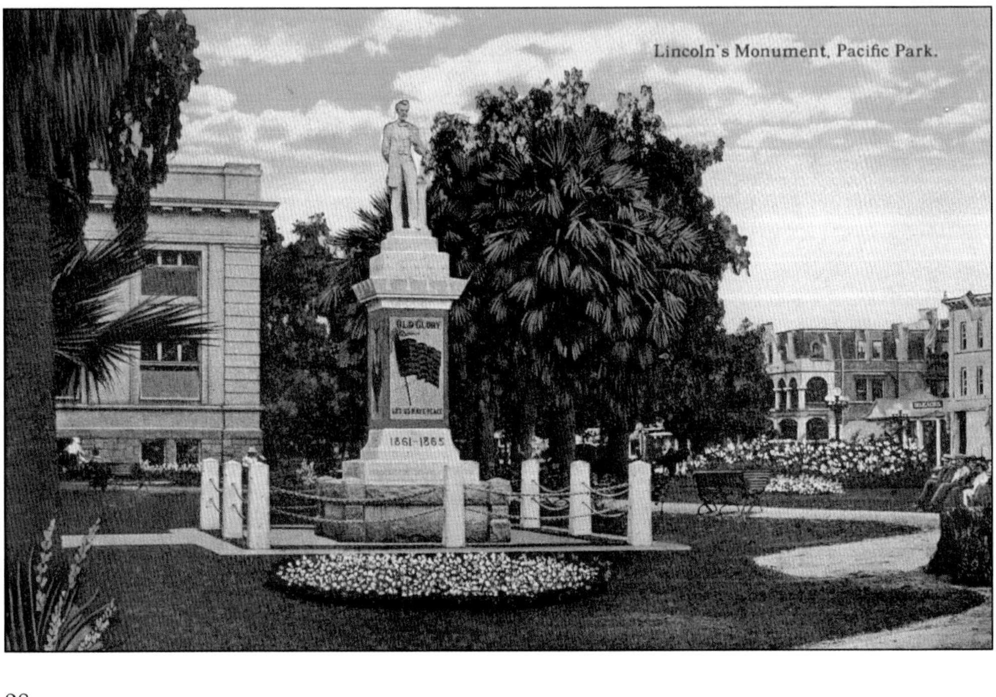

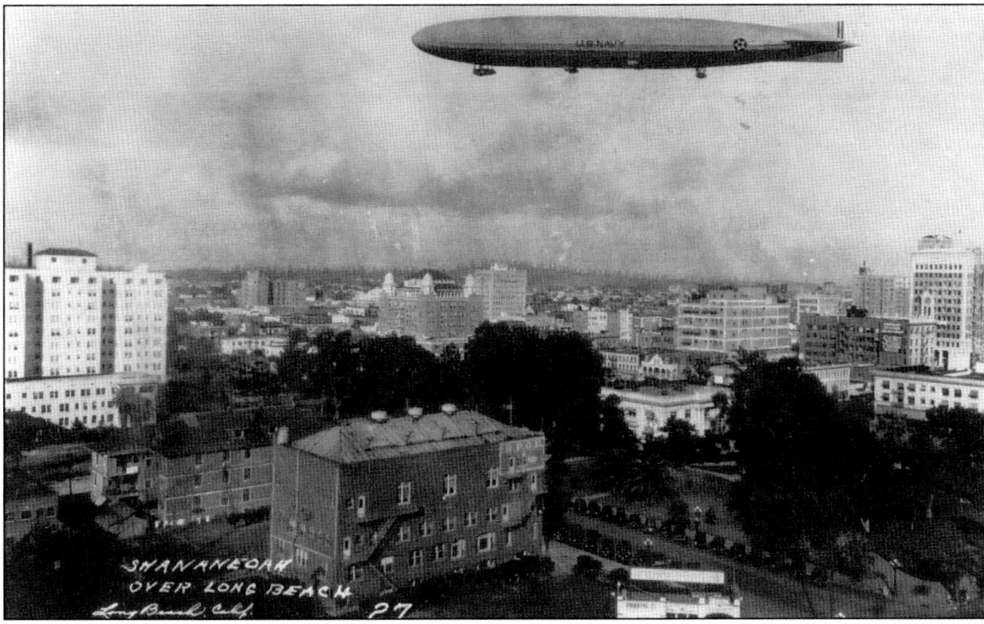

LONG BEACH CROWDS. Long Beach residents always turned out in large numbers to view military and civilian aircraft. In October 1924, residents saw the gigantic naval airship USS *Shenandoah* (ZR-1), above, pass over their beaches and buildings. The airship was on a maintenance and operations flight from its base at Lakehurst Naval Air Station in New Jersey to California and then to Washington. Airships were used by the navy for reconnaissance. Pictured below, residents visited the Long Beach Municipal Airport to view U.S. Army airplanes on the field. By 1925, the navy announced its intention to build hangars on the Long Beach Municipal field, thereby making "the field the airport for the Pacific Fleet." (Above, LOC; below LBMAA.)

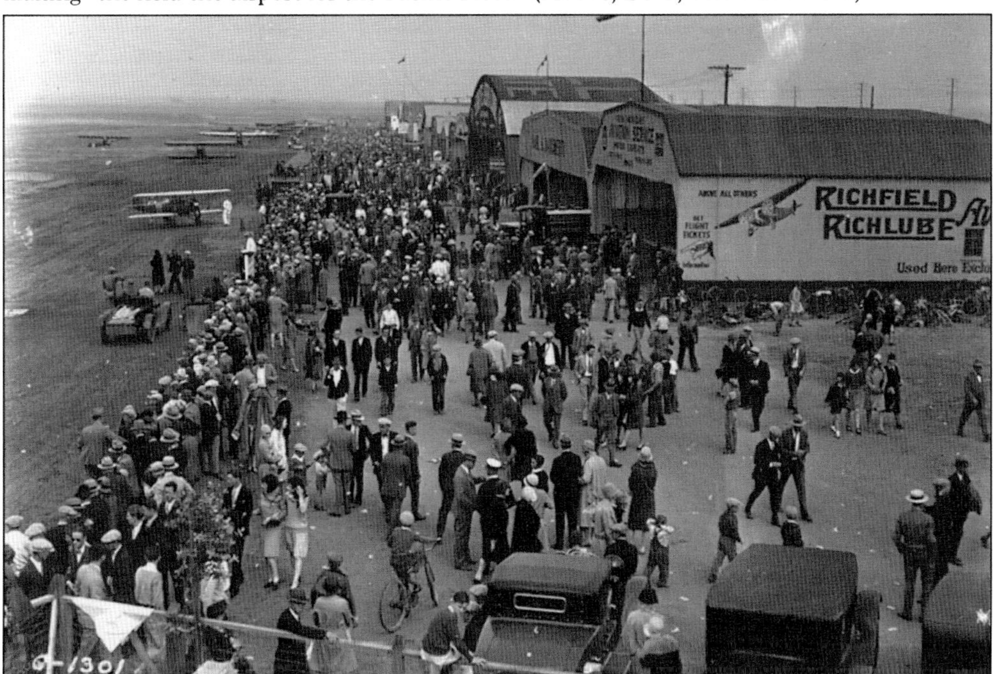

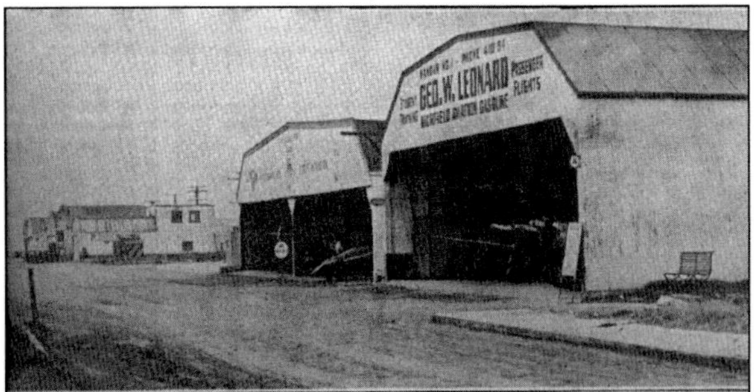

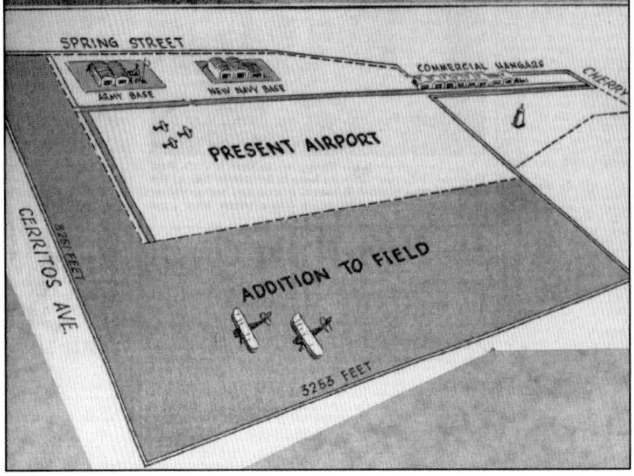

New Navy Base. In 1928, the city council approved funds to lengthen the runway more than 5,000 feet and construct a hangar, administration building, and a headquarters building for the U.S. Naval Reserve Aviation Base (NRAB) under the command of the Naval Operating Base at San Pedro. The LB-NRAB was officially commissioned on May 10, 1928. The extension of the runway was negotiated with the Montana Land Company, the owner of the property south of the airport. The Army Air Corps relocated from Clover Field in Santa Monica and can be viewed on Daugherty Field in the lower photograph.

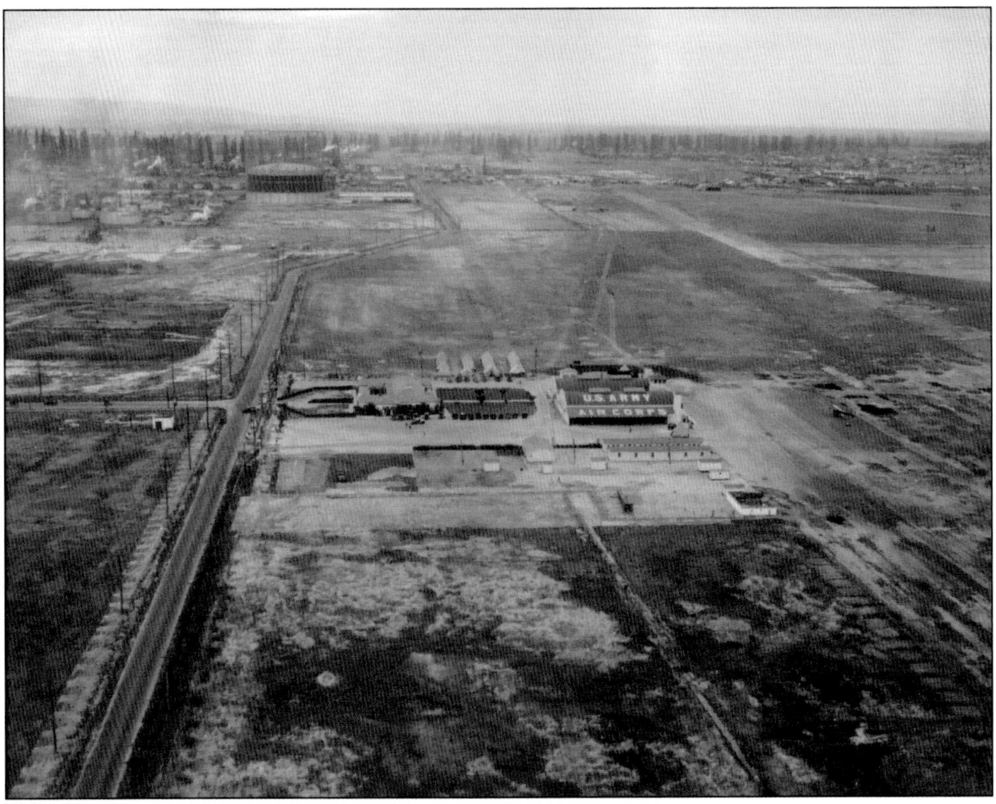

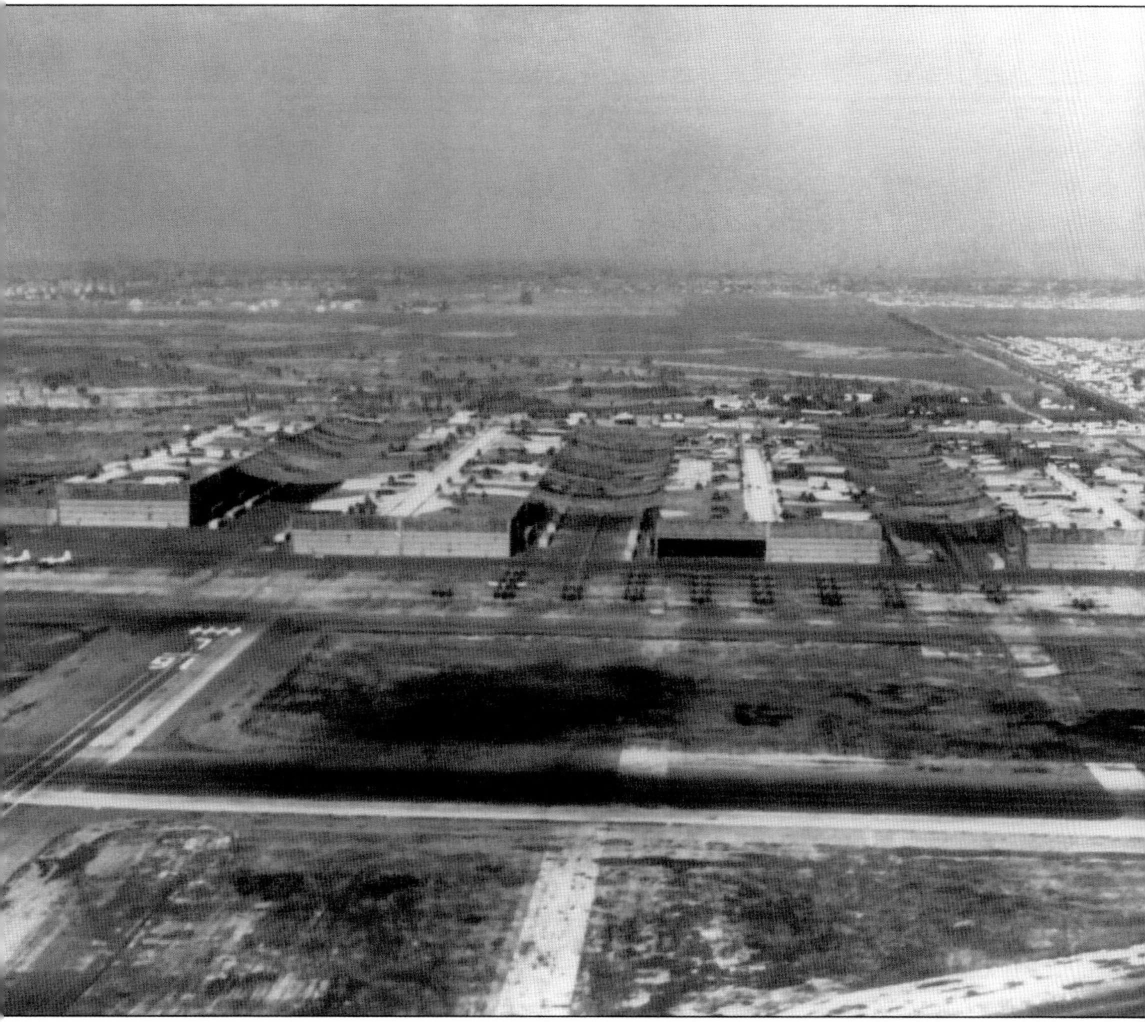

Aircraft Manufacturing Center. One of the largest land transfers in Southern California occurred when Janss Investment Company of Los Angeles purchased 8,000 acres of northeast Long Beach owned by the Montana Land Company. Of the purchased land, 1,400 acres was designated for development of a "big aircraft manufacturing center in the vicinity of the Long Beach Municipal Airport." A spokesperson for Janss explained to the *Los Angeles Times* in January 27, 1929, "The land was sought by Janss because of its excellent rail facilities, nearness to the harbor and the airport established it as the outstanding location. It is large enough to provide the manufacturer with adequate land for operations. It will enable us to establish a residential community for workers." In 1940, the area became Douglas Aircraft (pictured during the war under camouflage). Douglas had served as chief civilian aeronautical engineer for the U.S. Army Signal Corps. Douglas Aircraft produced the C-47 Skytrain, SBD dive-bomber, C-54 transport, A-20 and A-26 attack bombers, and B-17 bomber. During the war, 45 percent of the employees of Douglas Aircraft were female. The women were referred to as "Rosie the Riveters."

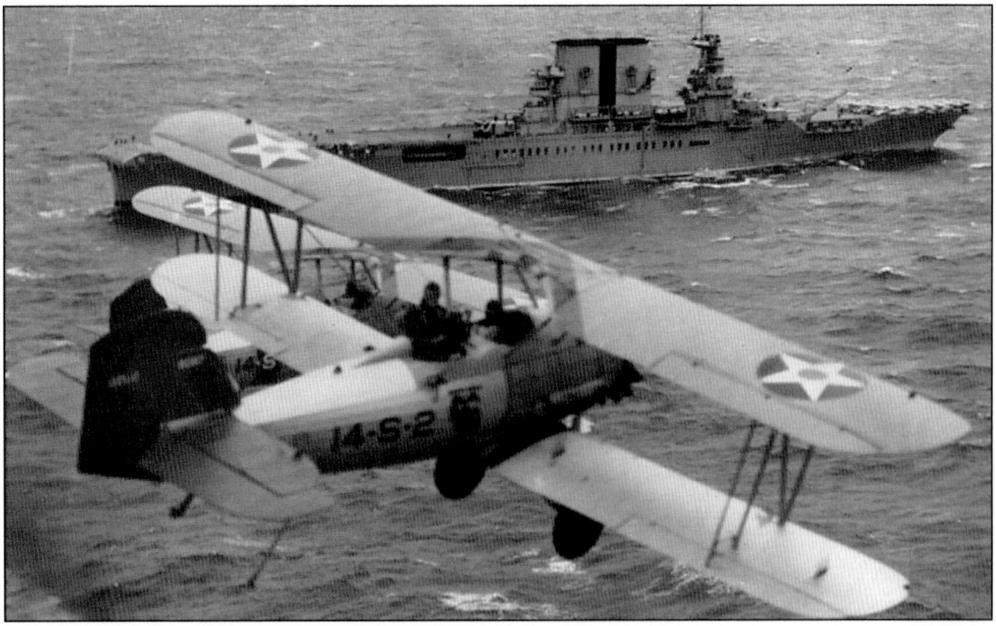

SARATOGA AND LEXINGTON. In 1928, the navy dispatched its two new aircraft carriers, the USS *Saratoga* (above) and the USS *Lexington* with 200 planes on board to anchor off the coast of Long Beach. Nearby Terminal Island (which included Allen Airfield) and the adjacent San Pedro harbor were sites for both army and navy aircraft and the first submarine base and yard on the Pacific Coast. The navy plane, a Vought, arrived on Daugherty Field in June 1928. While Long Beach welcomed the military presence at the airport, city officials made it clear they preferred commercial aviation. As a result, the city refused to sell airport property to the military. The navy thereupon purchased property offered by Susanna Bixby Bryant and relocated its air station to Los Alamitos by 1942. The navy left a small naval auxiliary air station at the airport until after the war ended but downgraded it from a naval base. (Both, USNHC.)

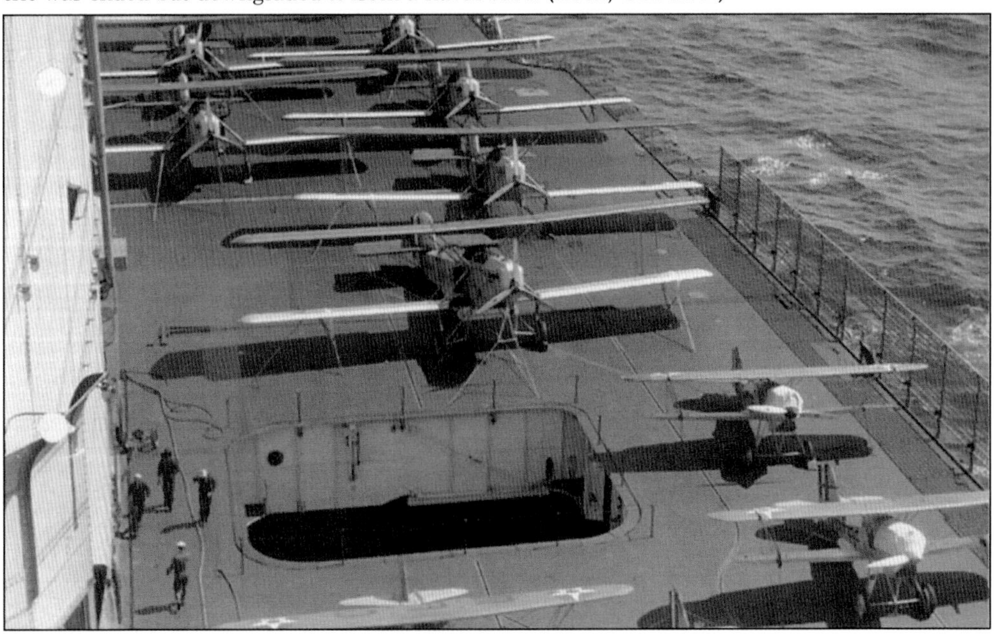

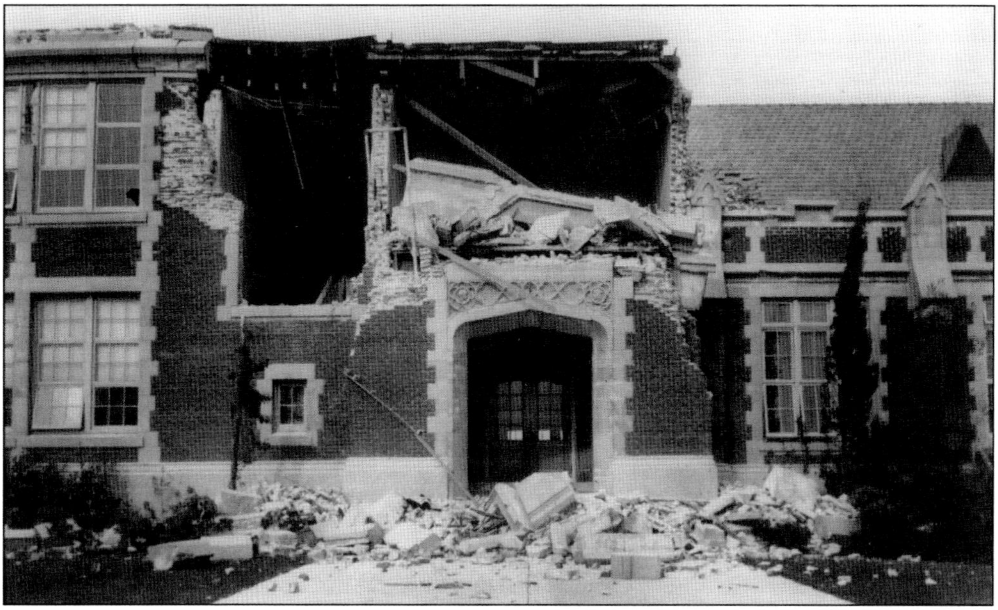

The Navy Helps after the Earthquake. The Great Depression hit Long Beach in the 1930s, as did the 6.3-magnitude earthquake on March 10, 1933. One hundred and twenty people died, including 5 children who died in a school gymnasium. There was $50 million in property damage. Ninety percent of the city's schools (John Muir School is pictured) were damaged and could not be occupied for over a year. School was conducted in city parks. Hundreds of residents lived in tents. More than 2,000 sailors and marines assigned to ships anchored off Long Beach helped clean up rubble. The Long Beach Municipal Airport was not damaged. State legislation was quickly passed requiring buildings to be built to stricter specifications, and many Long Beach buildings were rebuilt incorporating the new requirements and the latest in architectural design. (Above, U.S. Geological Service; below, AC.)

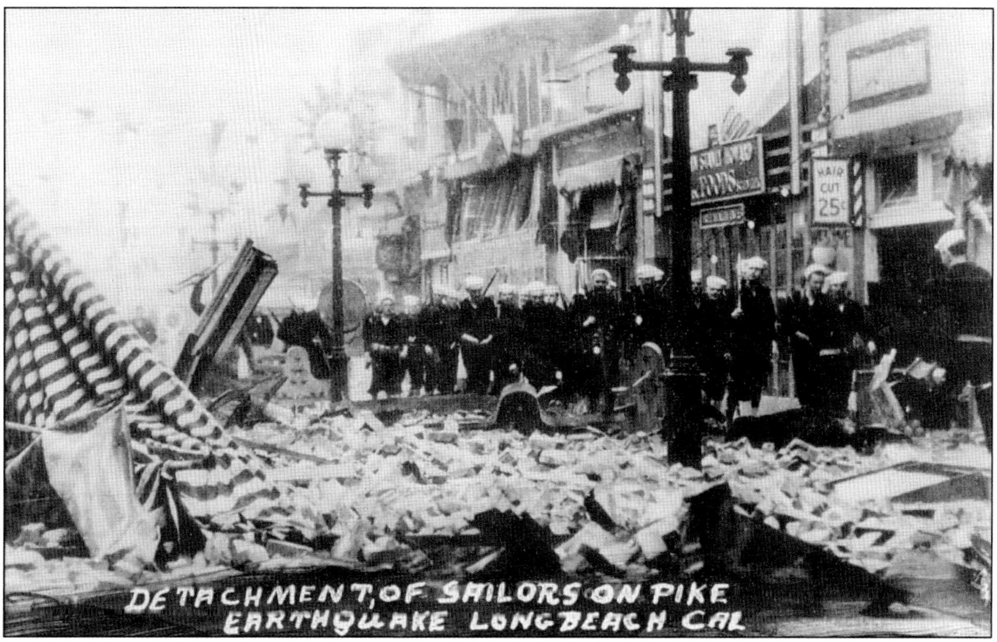

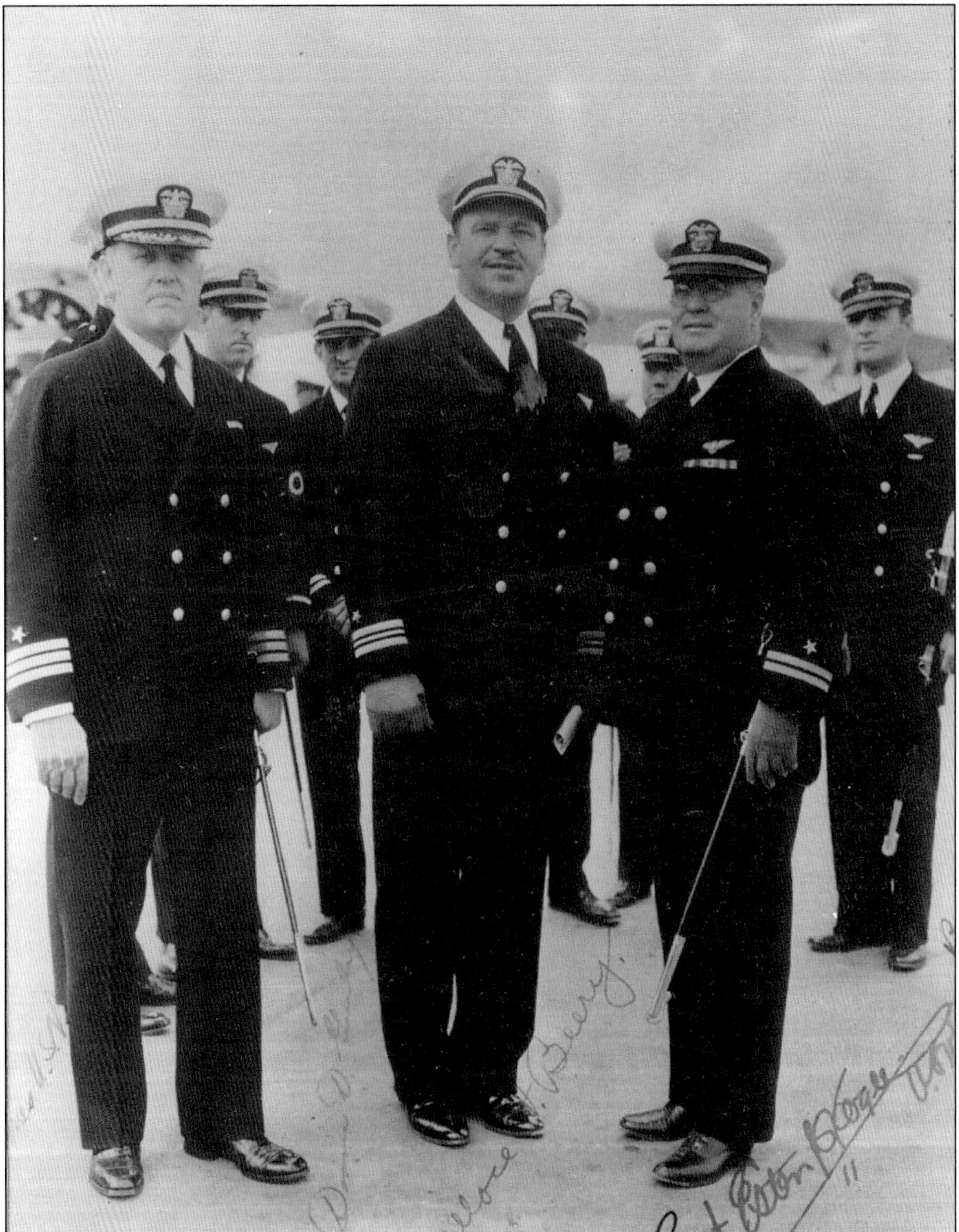

WINGS OVER THE NATION. In 1938, the city began its "Wings Over the Nation," a four-day event at the airport commemorating the anniversary of the Wright brothers' flight. In 1940, the event brought pilots and stewardesses from major airlines: Braniff, American Airlines, Delta, Transcontinental, Western Air, United Airlines, Western Air Express, Pennsylvania-Central Airlines, Chicago Airlines, and Southern Airlines. Actor and naval reserve officer Wallace Berry (pictured in the center) served as the toastmaster for the Long Beach Aviation Dinner, which capped the event.

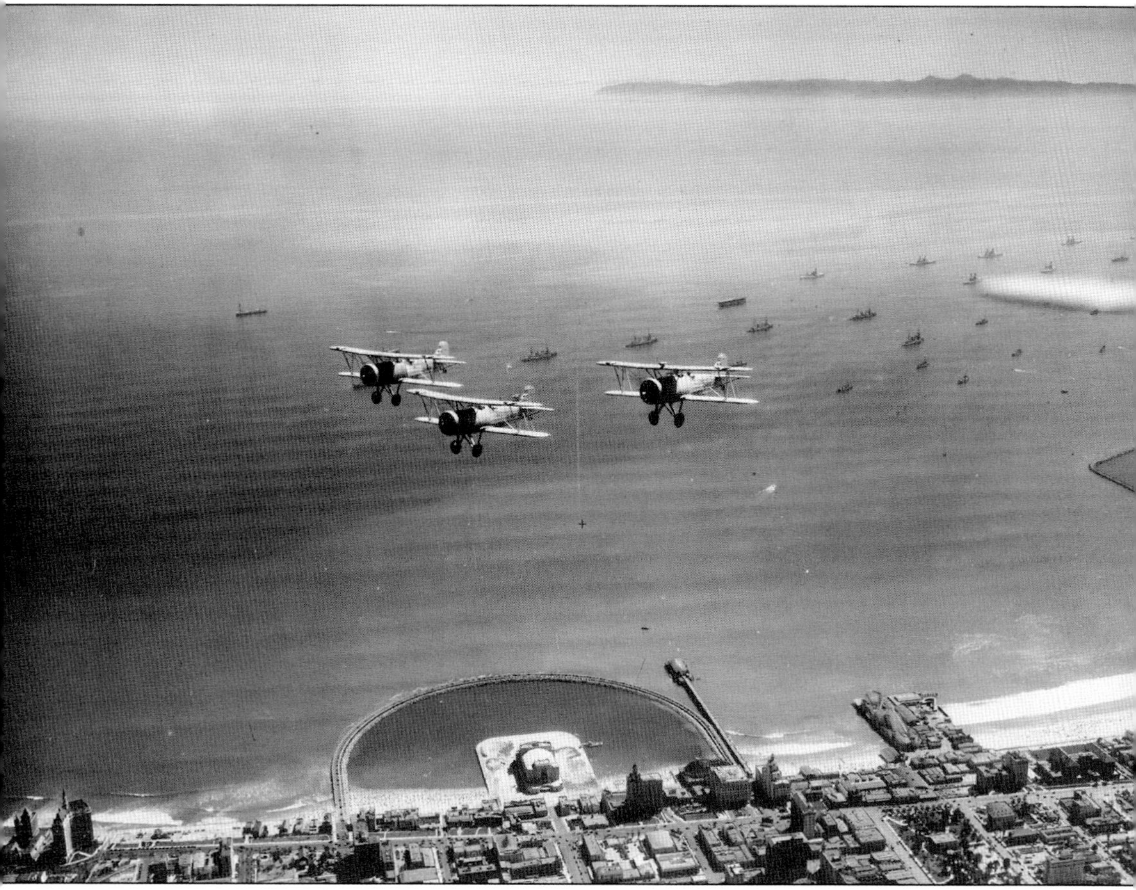

VN-13 RD 11 USNR. The configuration of Daugherty Field and Earl Daugherty's connections with the navy were the major reasons the navy established the NRAB and eventually assigned the VN-13 RD 11 Squadron, of which Daugherty was a member before his death. He was commissioned as a lieutenant, junior grade. The photograph shows the Naval Reserve Curtiss 020-1 "Helldrivers" of the LB-NRAB flying over downtown Long Beach. (RTG.)

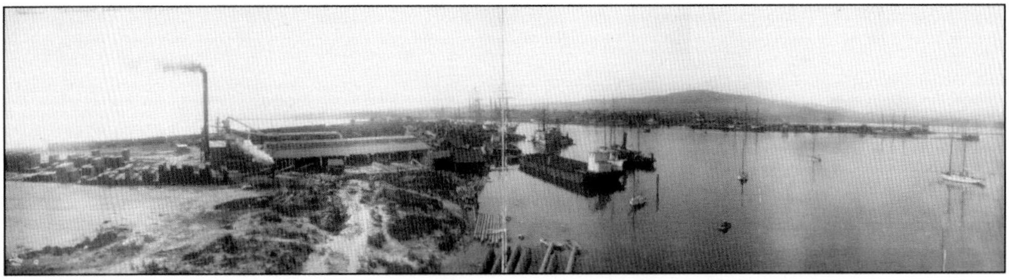

Terminal Island. The 1929 photograph above provides a glimpse of the sand-filled island between Long Beach and San Pedro, known as Rattlesnake Island because of the snakes that appeared after the area's frequent flooding. Later named Terminal Island by the Los Angeles Terminal Railroad, the island's ownership was divided between Los Angeles and Long Beach. It became the site for an airfield that provided support to the seaplanes for the battleships and cruisers anchored nearby and the submarine base in San Pedro harbor. Three asphalt runways were eventually developed, with the longest at 4,900 feet. In 1938, the air station was commissioned as NAS San Pedro. The airfield, initially named Allen Field, was changed by the navy to Reeves Field. The navy operated a Naval Air Reserve Training Facility until 1942, when it was relocated to the new Los Alamitos Naval Air Station. NAS-TI closed in 1947. Hundreds of Japanese Americans were residents of Terminal Island, many of them owners of fishing vessels. They were rounded up by federal order in 1942 and processed at the Terminal Island federal prison that had been built in 1938. They had their property confiscated and were relocated to federal internment camps. (Both, LOC.)

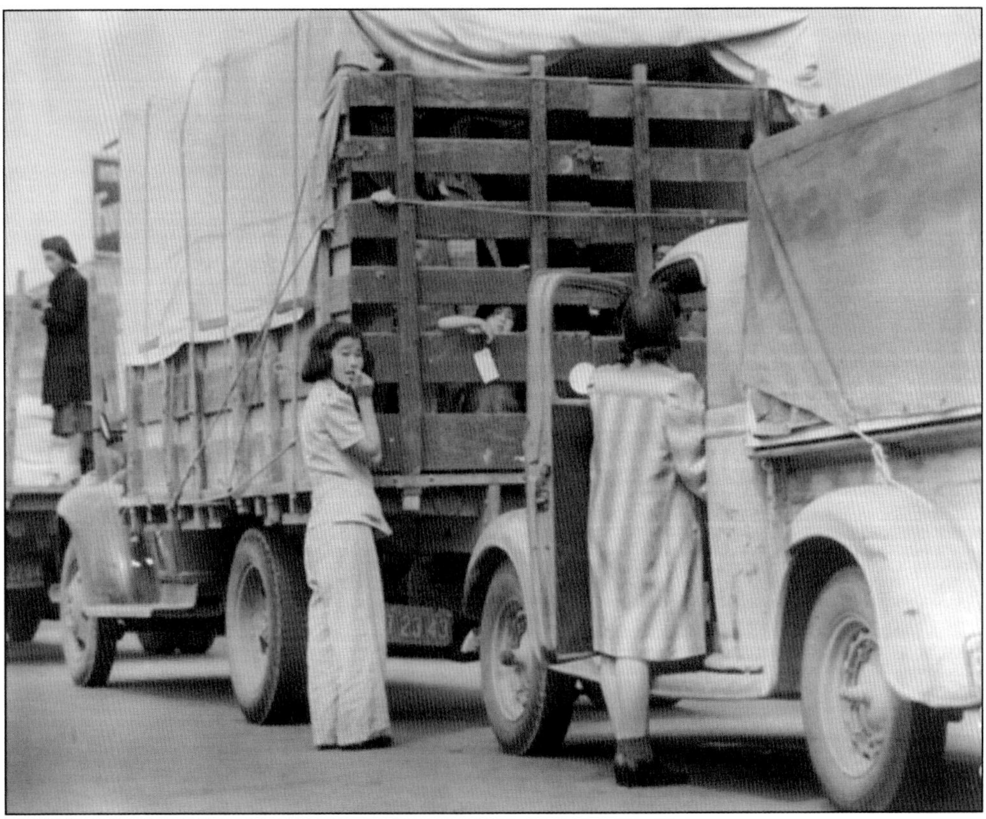

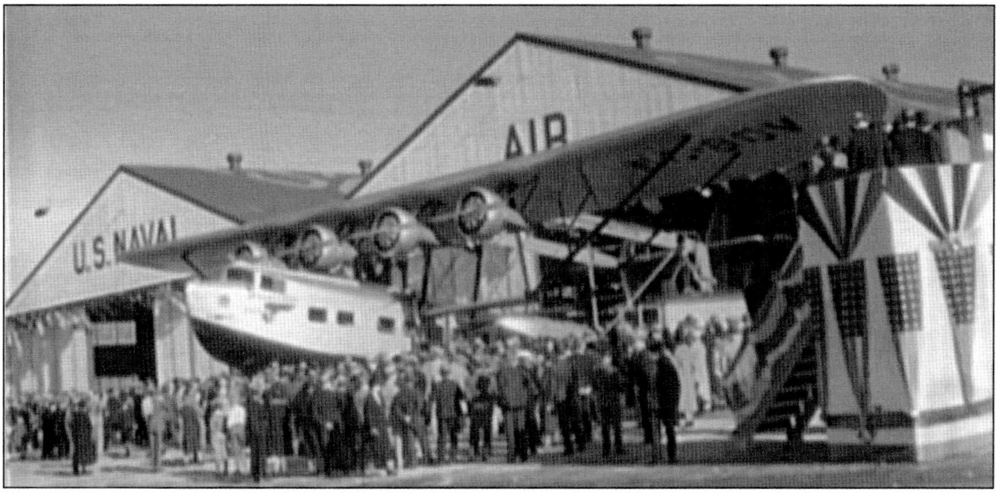

AMERICAN CLIPPER. While Long Beach was trying to entice the navy to build a base on its 40 acres of Terminal Island in 1939, the Los Angeles Harbor Board gave Pan American Airlines rights on its portion for a "southland base" for its giant seaplane clipper. Pan Am christened its giant Boeing 314 Seaplane *American Clipper* at Terminal Island and started service to New Zealand in 1939. The *Clipper* was the largest commercial airplane for its time and called a "flying boat." Terminal Island became a major port of call for the Pan American clipper line. In 1942, the *American* and other "clippers" were sold to the navy. (LOC.)

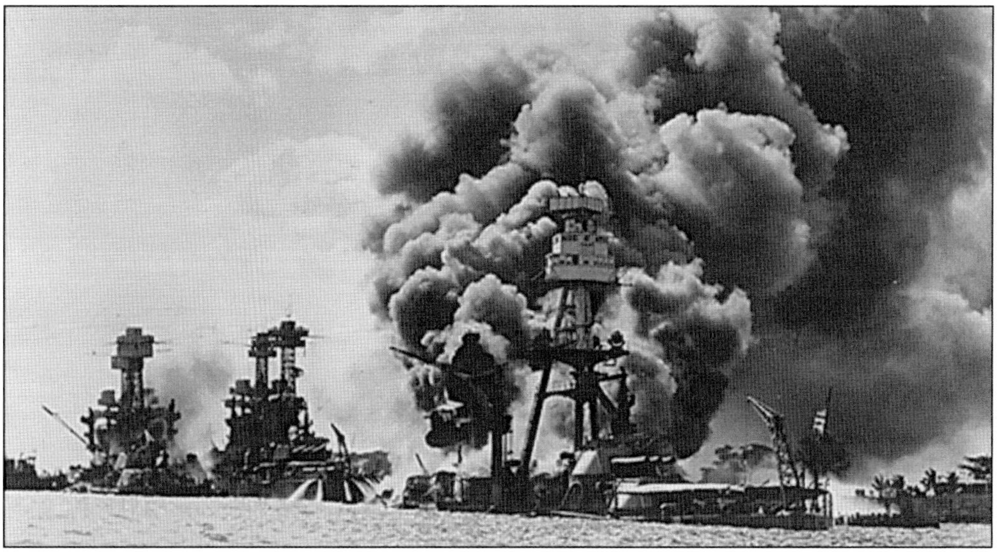

LONG BEACH SHIPS IN PEARL HARBOR. Many ships anchored off the coast of Long Beach during the late 1930s were destroyed or severely damaged during the December 7, 1941, attack on Pearl Harbor. These ships included the *West Virginia*, *Maryland*, *California*, *Oklahoma*, *Tennessee*, *Nevada*, and *Arizona*. USS *West Virginia* (left) was severely damaged. USS *Tennessee* was damaged and the USS *Arizona* sunk. Before Pearl Harbor, the City of Long Beach had negotiated with the army to lease space for $35,000 a year at the airport. The city also offered a lease of 73 acres for $1 a year if the army would clean up an area used by the city as a dump. The army wanted Long Beach as the location for its "ferry command" to oversee the assembly of military airplanes manufactured by seven Southern California airplane manufacturers. In August 1941, the 6th Ferrying Group of the Army's Air Transport Command was assigned to Long Beach Municipal Airport. (NA.)

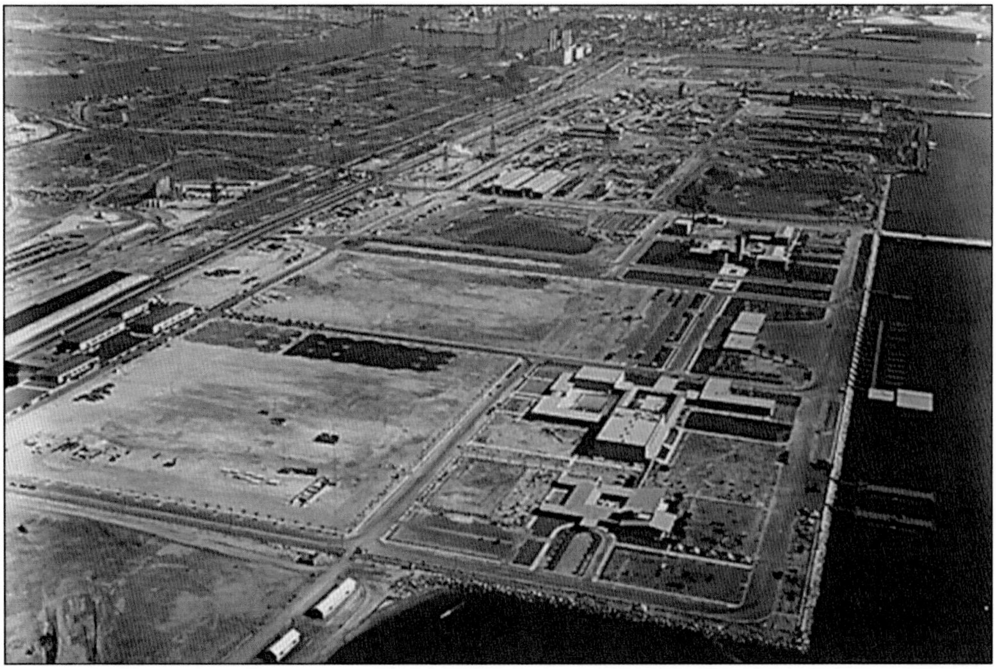

Roosevelt Base, Terminal Island. The proximity of a large airfield, shipyards, and a breakwater convinced the navy to build the U.S. Naval Dry Docks and Roosevelt Base on the south side of Terminal Island in 1943. Military aviation operations continued on Terminal Island as planes assembled at the nearby aircraft plants of Douglas, Vultee, and Lockheed were flown to NAS Terminal Island and checked out for delivery to the war front (shown). The shipyard facility was renamed the Terminal Island Naval Shipyard in 1945 and in 1948 became the Long Beach Naval Shipyard (NSY). Roosevelt Base was designed by Paul Revere Williams, the first African American licensed architect. After the base closed in 1997, historic preservationists attempted to save the buildings, but the entire site was demolished and used as a commercial shipping area for the Port of Long Beach. (Both, NA.)

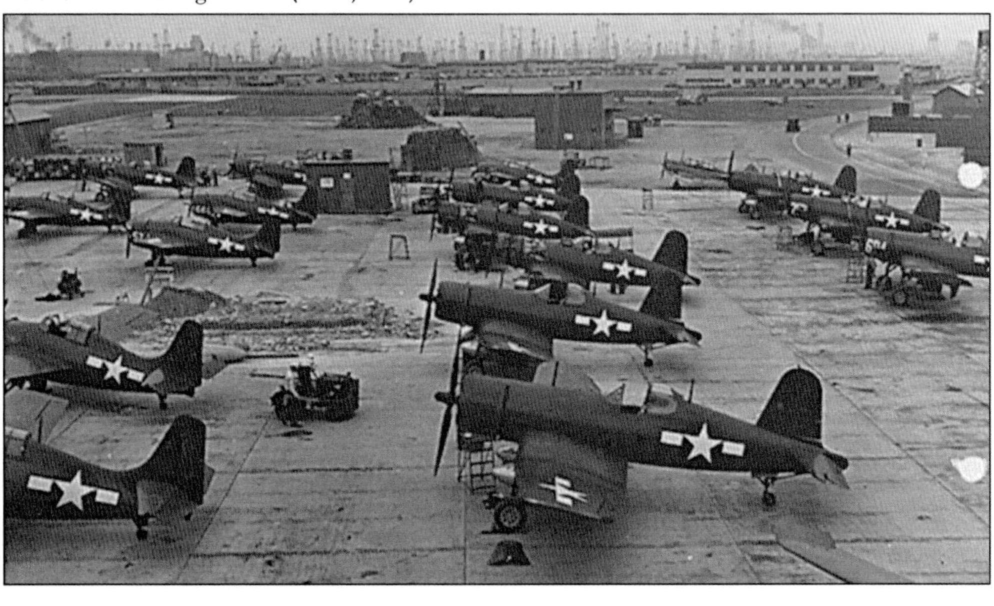

Eight

Women Aviators in Long Beach

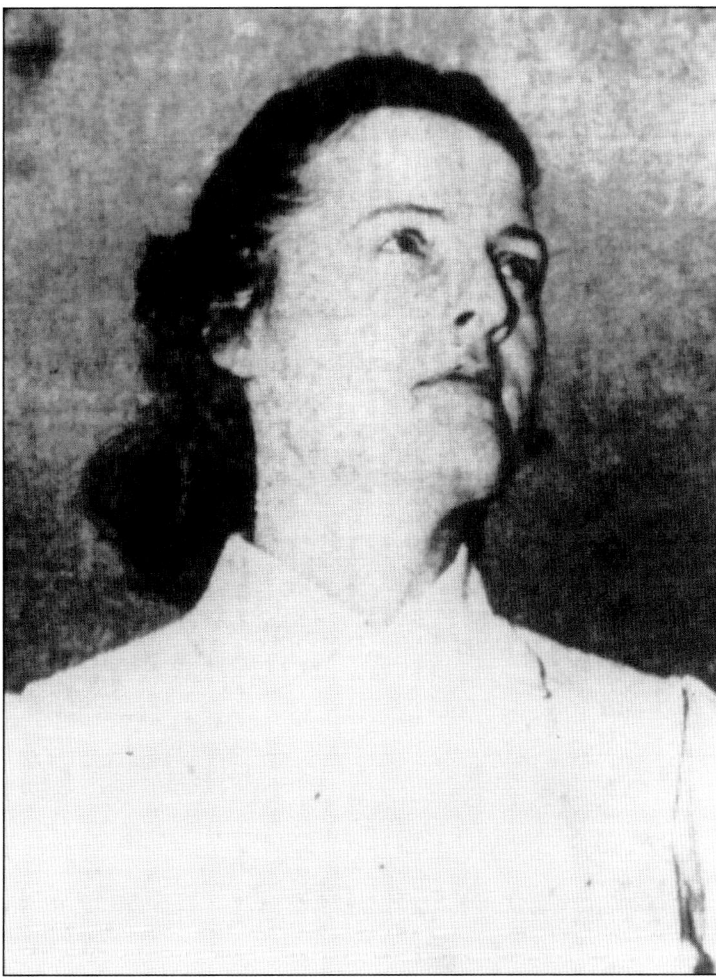

Gladys O'Donnell. From its earliest days, Long Beach was a major site for women in aviation. After John Montijo taught Amelia Earhart how to perform aerobatics, she flew stunts in Earl Daugherty's air rodeo. Faster planes and the promise of cash prizes and instant fame attracted both men and women to become involved in aviation, including Gladys O'Donnell, the first licensed female pilot in Long Beach.

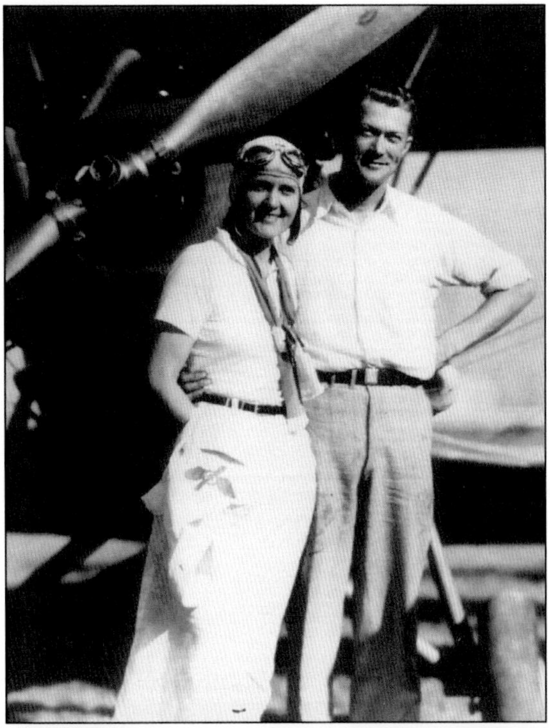

GLADYS AND LLOYD O'DONNELL. Gladys (Livingston Berry) O'Donnell was born in 1904 and married car salesman Lloyd O'Donnell when she was 18. Lloyd opened the O'Donnell School of Aviation at Daugherty Field and owned the first plane manufactured by Claude Ryan. Gladys was initially afraid to fly, but Lloyd persuaded her. After only 40 hours of flying, Gladys defeated Amelia Earhart in the 1929 eight-day transcontinental Women's National Air Derby from Santa Monica to Cleveland. The race attracted 20 women, including Evelyn "Bobbi" Trout and Pancho Barnes. Louise Thaden won the race. O'Donnell came in second, winning $1,950; Earhart came in third. The O'Donnell School of Aviation was operated by Lloyd O'Donnell at Daugherty Field, who set his own aviation record in 1929 when successfully completing the first glider tow.

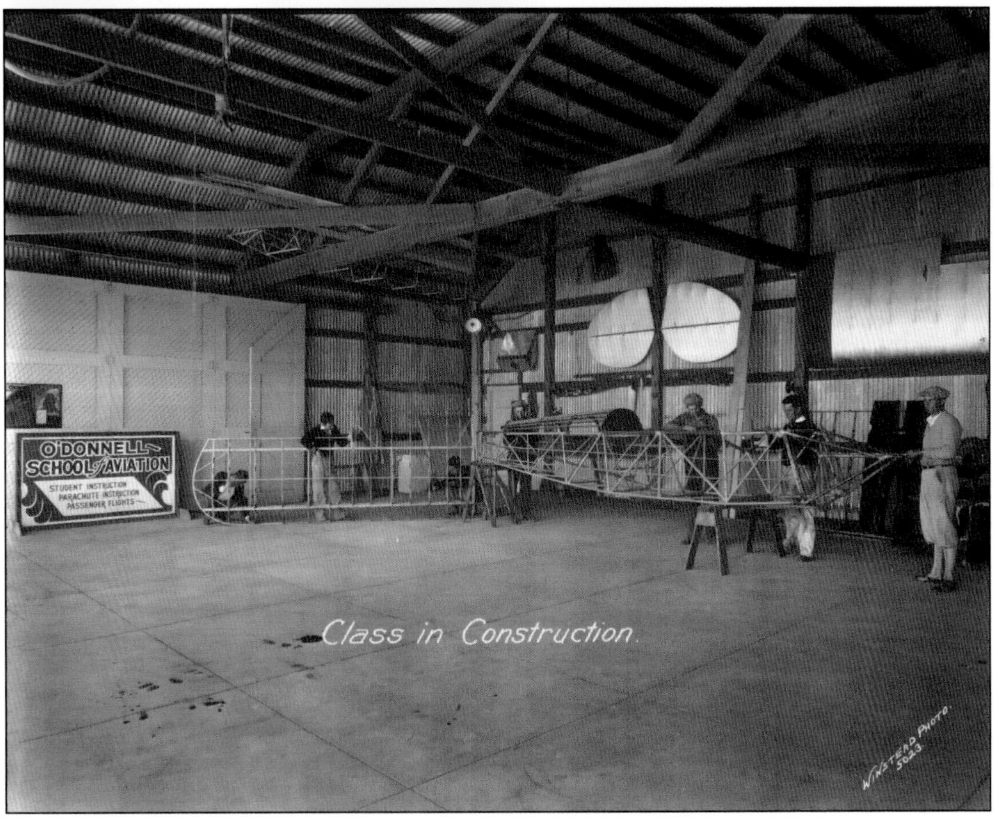

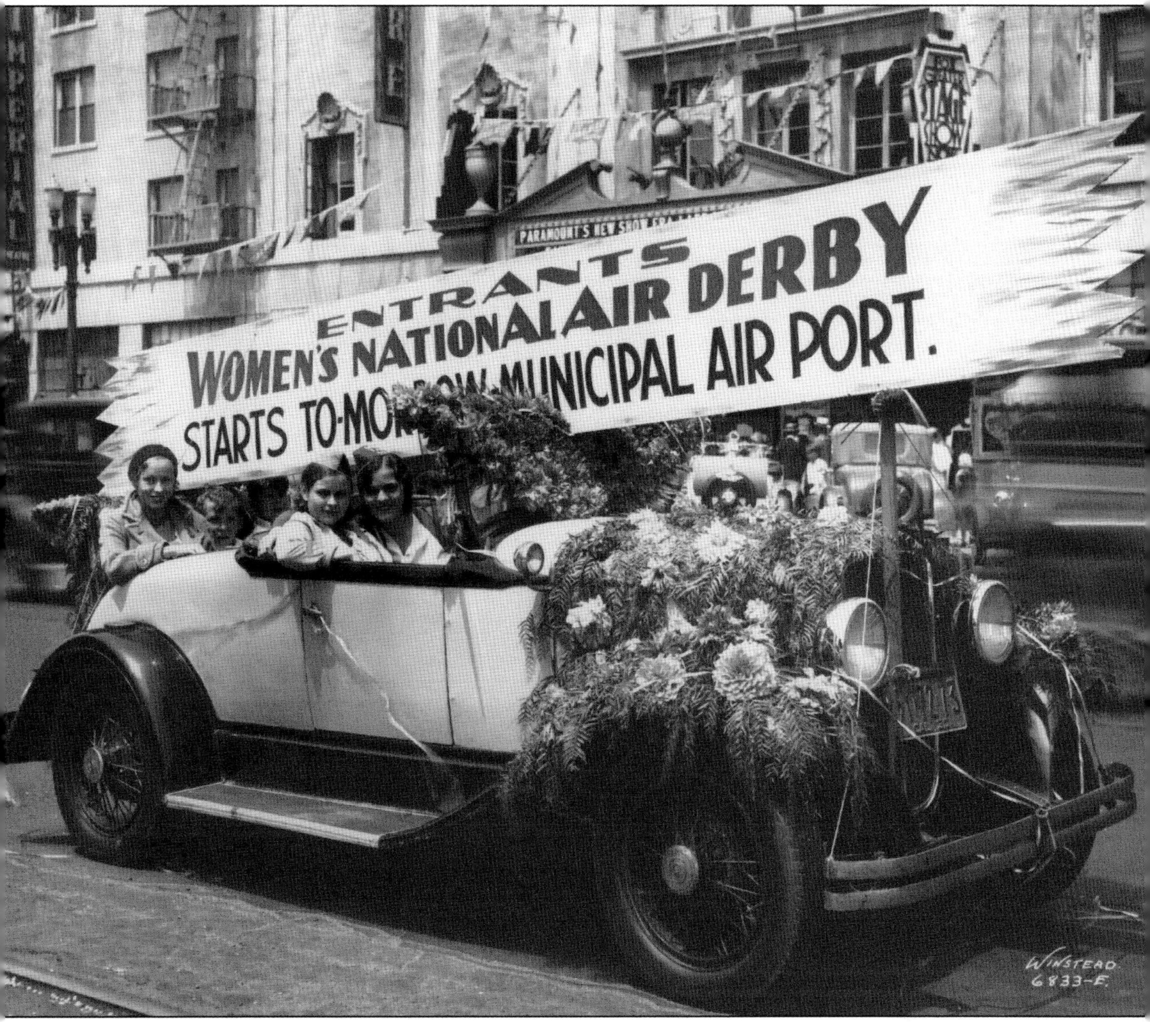

Powder Puff Derby. Gladys O'Donnell was called "the flying housewife" with her win at the 1930 Women's Air Derby, which started at Long Beach and ended in Chicago. O'Donnell flew a 1929 Waco Taperwing Biplane, sponsored by the *Press Telegram*. Humorist Will Rogers referred to the race as a "powder puff derby" because the flyers supposedly checked their makeup before taking off. To hold the Women's Air Derby, Long Beach removed the gas beacon light from the field. The airfield was officially dedicated to Earl S. Daugherty during the race. In 1929, Long Beach surged with the largest number of women qualifying for flying licenses: Mrs. W. R. Sweeley, Edna Coulter, Melba Gorby, Virginia Blume, and L. M. Partee. "Long Beach society woman" Mrs. C. C. McWhinney (who took lessons from Earl Daugherty) soloed in her Laird plane in 1928, only to crash and die on April 12, 1929.

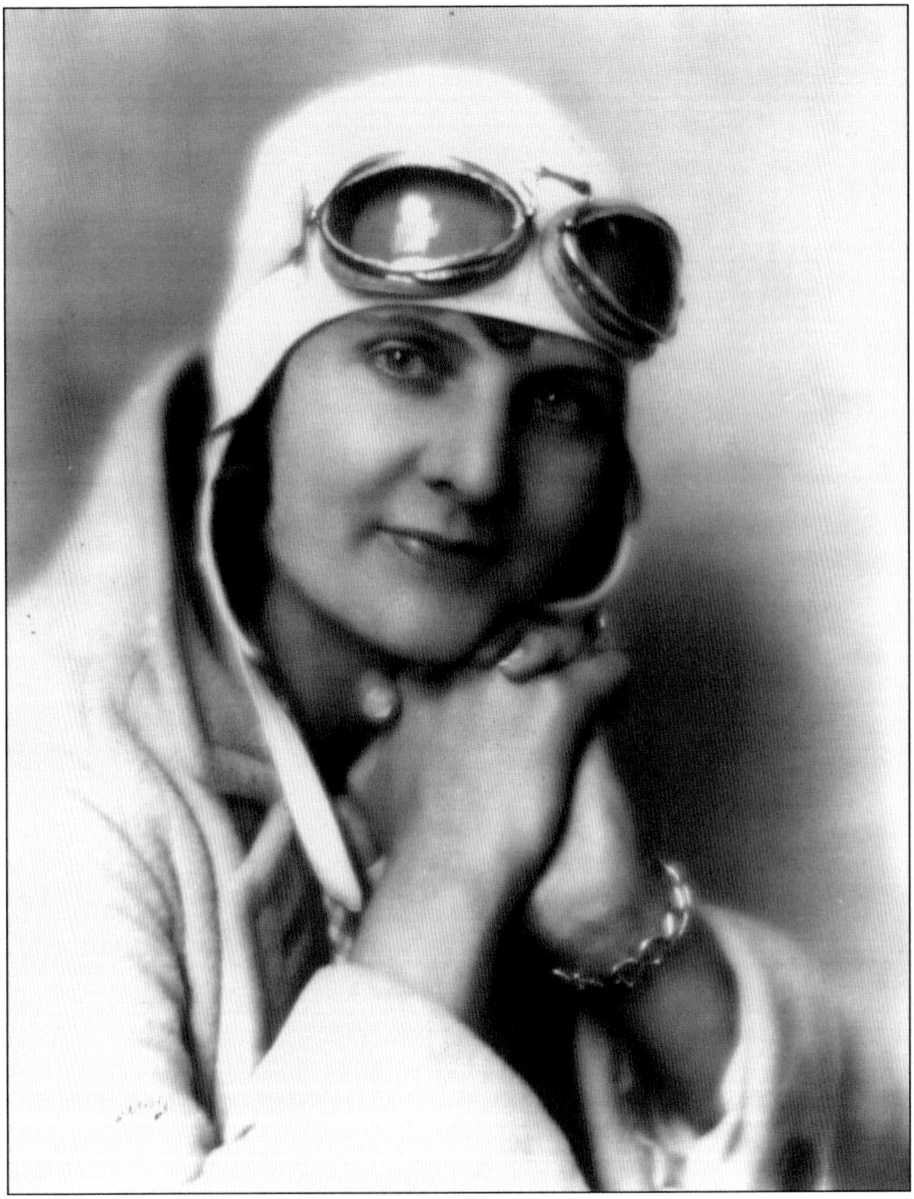

Ninety-nines. The first organization of women pilots was formed in 1929 and named "Ninety-nines" to signify the number of women who organized the group. Gladys O'Donnell (pictured), a charter member of the group (with Amelia Earhart, who often visited her in Long Beach), served as first "governor" of the Ninety-Nine Southwest area. O'Donnell did a radio program, *Sky Doings*, throughout the 1930s. She won the 1937 Amelia Earhart Memorial Race in Cleveland. An instructor for the Civil Aeronautics Authority, she served on the Long Beach Aviation Commission several terms. In 1941, she and her husband, Lloyd, operated Sequoia Field in Visalia, California, training pilots for military flying. They also owned O'Donnell Oil Company, and she cofounded Hydro-Test Tubing Service. In 1967, Gladys defeated Phyllis Schafly for president of the National Federation of Republican Women and served in various positions in the Republican party until her death in 1973. Long Beach recognized her contributions to aviation in 2006 with the naming of a fire department fire engine "Pioneer."

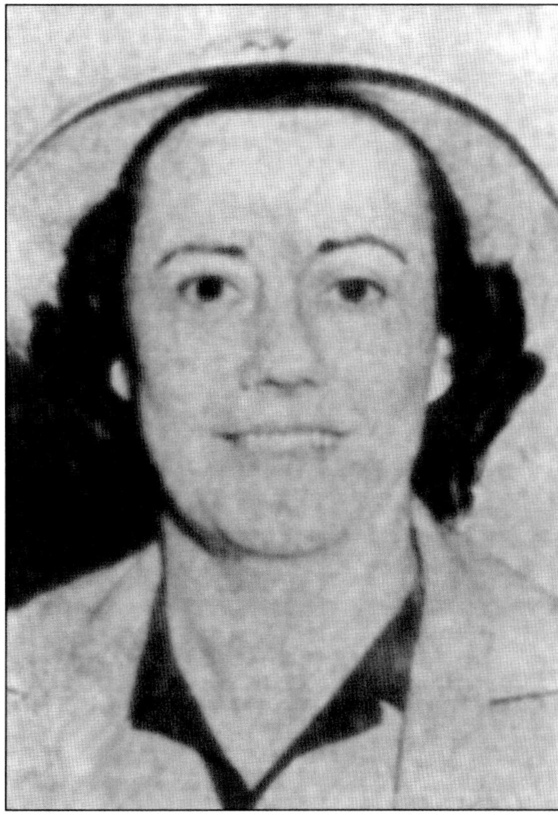

Kay Daugherty. Earl S. Daugherty's widow, Kay, remained active in aviation for many years. Her husband taught her how to fly at the Willow field in the 1920s. During the war, she served as a first lieutenant and executive officer of the Civil Air Patrol Squadron (pictured below) and helped form a cadet training program in Long Beach. She remained active in aviation, managing the Daugherty Flying Service. She lived until the age of 101, residing on Earl Street in Long Beach.

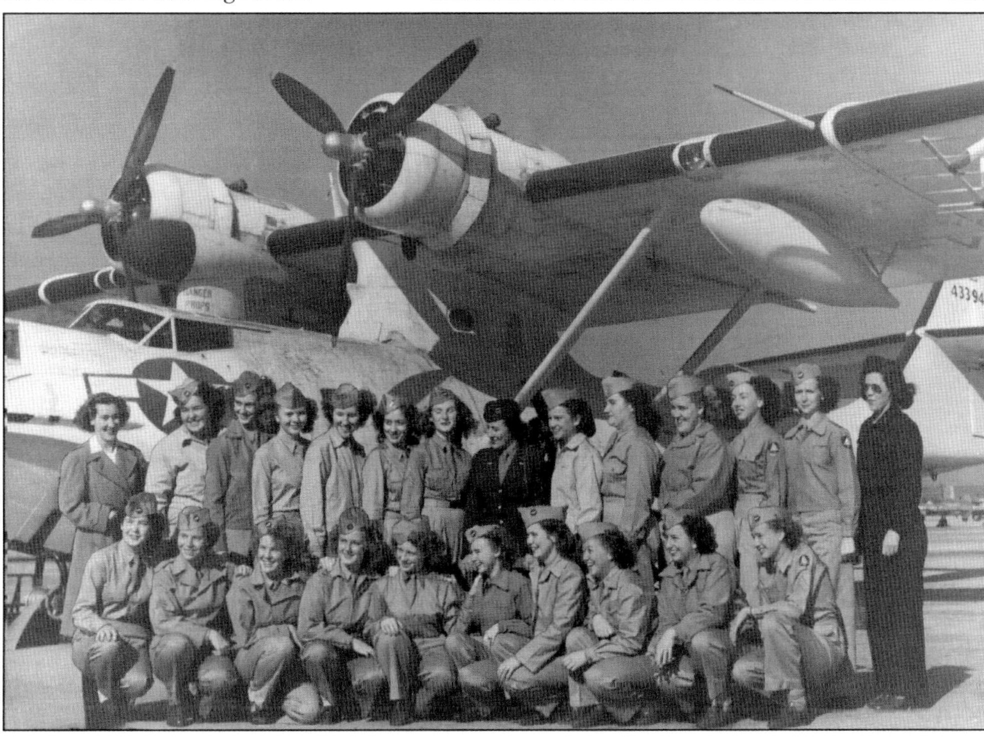

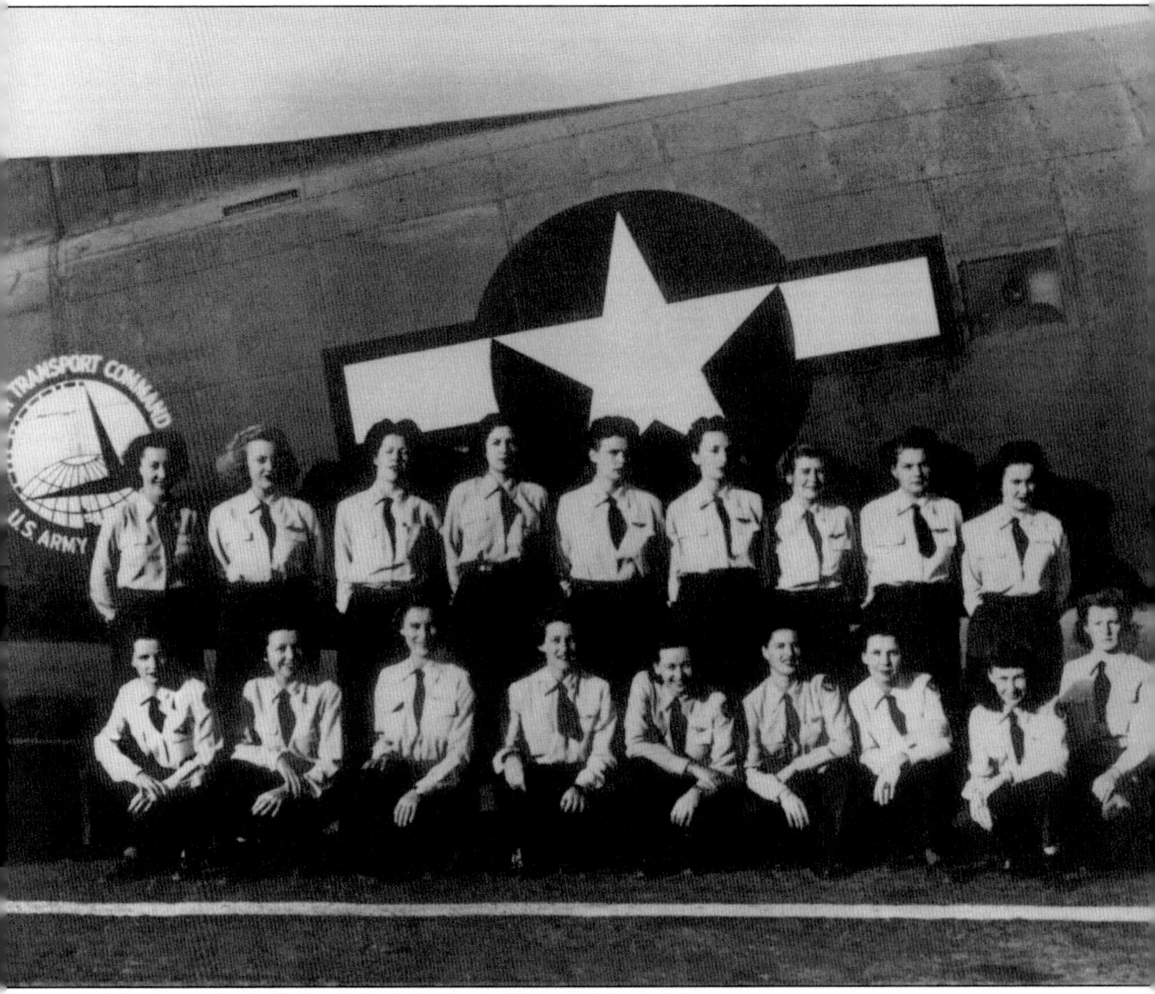

WAFS AND WASP. Daugherty Field and Reeves Field were the center of ferrying military aircraft produced in Southern California. Women were utilized to fly the thousands of the planes assembled at Douglas Aircraft, Vultee, North American, and Lockheed to military bases. Though 25,000 women applied, only 1,100 graduated and served between 1941 and 1944 in the Women's Auxiliary Ferry Squadron (WAFS) and the Women Airforce Service Pilots (WASP). The programs became the WASP in 1943. They trained in military flying at Avenger Field, Sweetwater, Texas. WASP in Long Beach were attached to the 6th Ferrying Group Air Transport Command, U.S. Army Air Corps. Jeanne Robertson identified some of the women who served in Long Beach: from left to right, (first row) Rena Wilkes, Lauretta Beatty, Iris Cummings, Lewise Coleman, Dorothy Webb, Jean Landis, Dorothy Kocher, Ruth Thompson, and Helen Richards; (second row) Nadine Ramsey, Betty Tackaberry, Katherine Loft, Evelyn Loft, Evelyn Trammell, Thelma Farris, Deborah Truax, Virginia Hill, Carol Fillmore, and Barbara Erickson London. Two hundred WAVES (Women Accepted in Voluntary Emergency Service) and SPARS (Semper Paratus—Always Ready, Coast Guard Women's Reserve) served on Terminal Island as part of the Naval Air Transport Service.

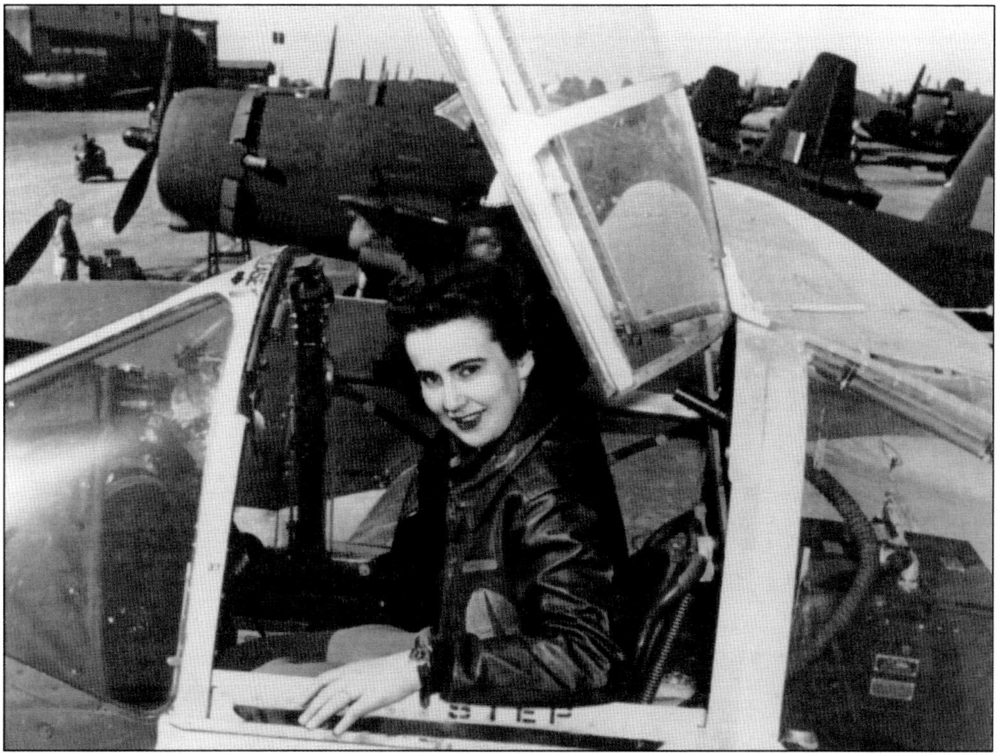

BARBARA (ERICKSON) LONDON. Barbara Jane Erickson joined the WAFS in 1942, was assigned to the 2nd Ferry Group in Wilmington, Delaware, and then led the women assigned to the 6th Ferrying Group at the Long Beach Municipal Airport–Daugherty Field. As squadron commander, Erickson was the only woman to receive the Air Medal for her distinguished service, having ferried a P-47, P-51, and C-47 over 2,000 miles in just five days. In 1948, she was commissioned a major in the Air Force Reserve. She married Jack London Jr., whom she met in Long Beach, and together they later operated United States Aviation, a flight school. The company and aircraft parts business closed during the Korean War. She remained involved in aviation and operated Barney Frazier Aircraft, Inc., and served as a board member of the "Powder Puff Derby" association. Long Beach acknowledged her contribution to aviation in 2006 with the dedication of Barbara London Drive at the Long Beach Municipal Airport.

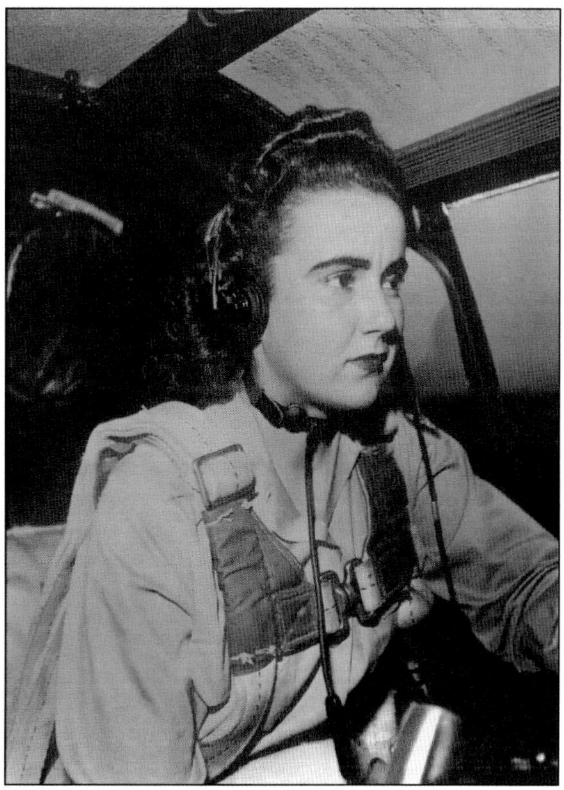

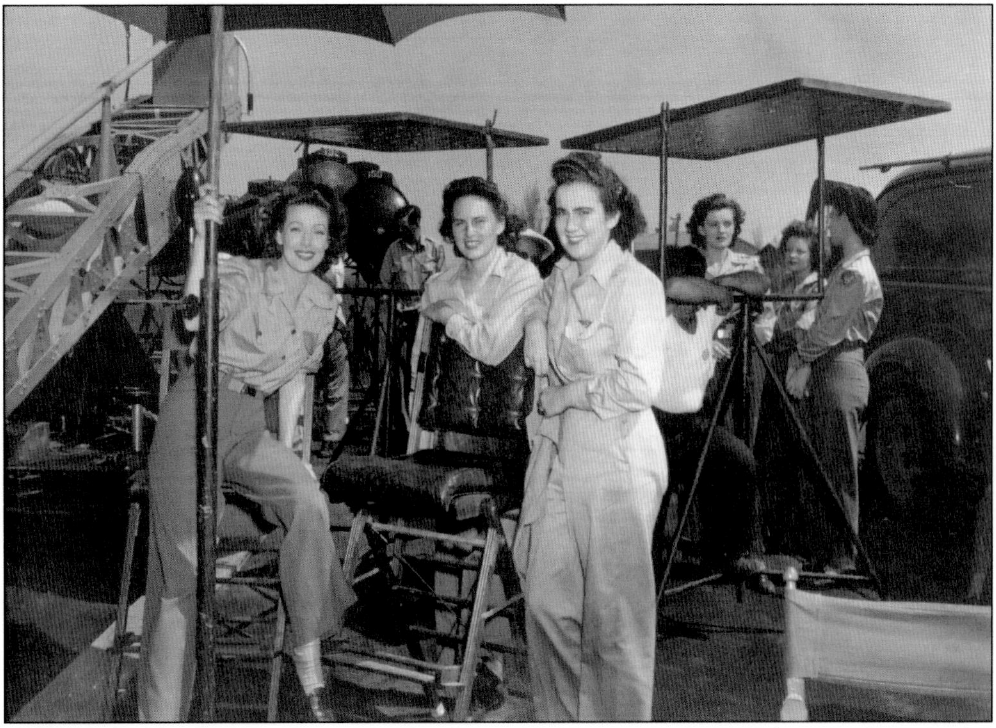

WASP and Hollywood. Actress Loretta Young (left) consults with Barbara Erickson (right) on the set of the 1943 movie *Ladies Courageous*, which glamorized the dangerous work WASP performed. Movie stars and officials visited Long Beach during the war, among them Bob Hope, Ruth Hussey, Kate Smith, and Eleanor and Pres. Franklin Roosevelt.

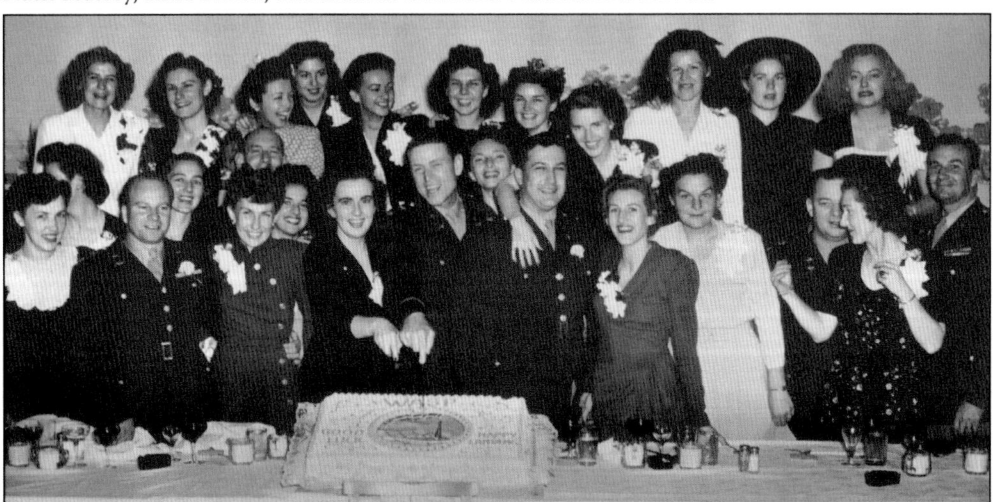

An End of an Era. WASP were deactivated on December 20, 1944, as a result of complaints from male aviators and politicians. They were denied veterans benefits and medical care for the injuries they suffered. Thirty-eight women lost their lives while ferrying planes. In 1971, Congress finally gave veteran status to WASP, and in 1979, they were finally given honorable discharges. On July 9, 2009, Pres. Barack Obama signed into law a bill conferring the Congressional Medal of Honor on all WASP.

Nine

The Long Beach Municipal Airport Terminal Building

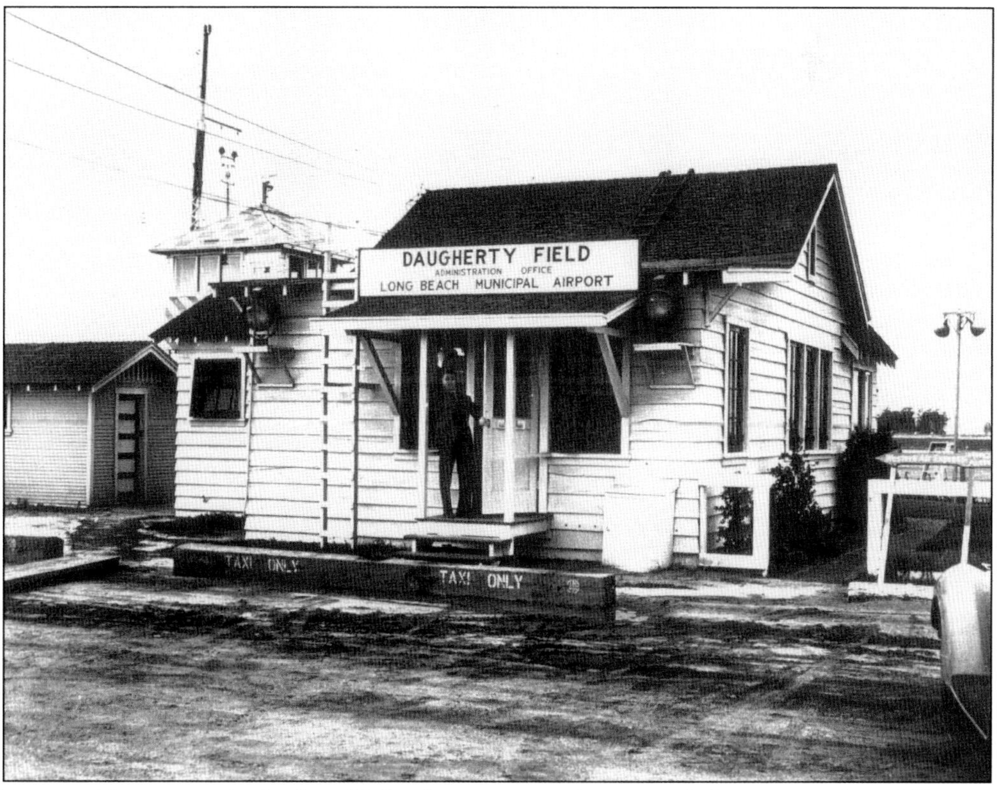

Lack of Facilities. In the late 1930s, the city council approved plans to purchase 255 acres adjoining the Long Beach Municipal Airport–Daugherty Field and to construct a three-story administration building and tower at the east side of the field at a cost of $200,000 in response to airline complaints about the small shack that served as the terminal and the lack of facilities on the field.

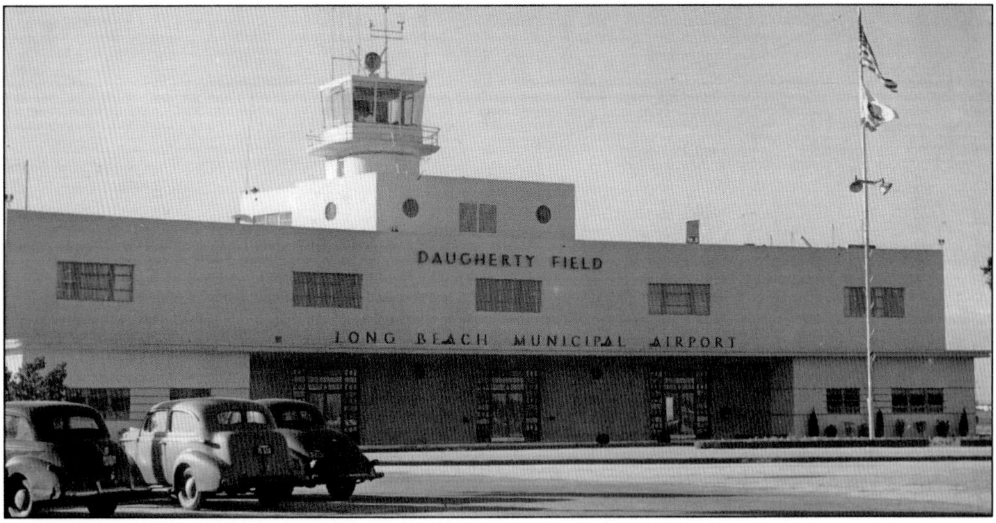

Streamline Moderne. Designed by W. Horace Austin and Kenneth S. Wing in the streamline moderne style of later art deco architecture, the terminal building looks more like a ship than an airport terminal. Its distinguishing characteristics include smooth walls, lack of ornamentation, flat roofs, railings, and porthole windows. Planned for expansion, it is shaped as a segment of an arc with a radius of 285 feet and a length of 170 feet. The third floor is set back, making the building 60 feet high. Opening ceremonies were cancelled because of Pearl Harbor. Instead, military guns and soldiers were billeted around the terminal building and in the basement for the duration of the war. Showers and hot water tanks were installed for the soldiers. Barracks were built adjacent to the terminal. The building was repainted in camouflage until 1945, when its pastel colors were returned.

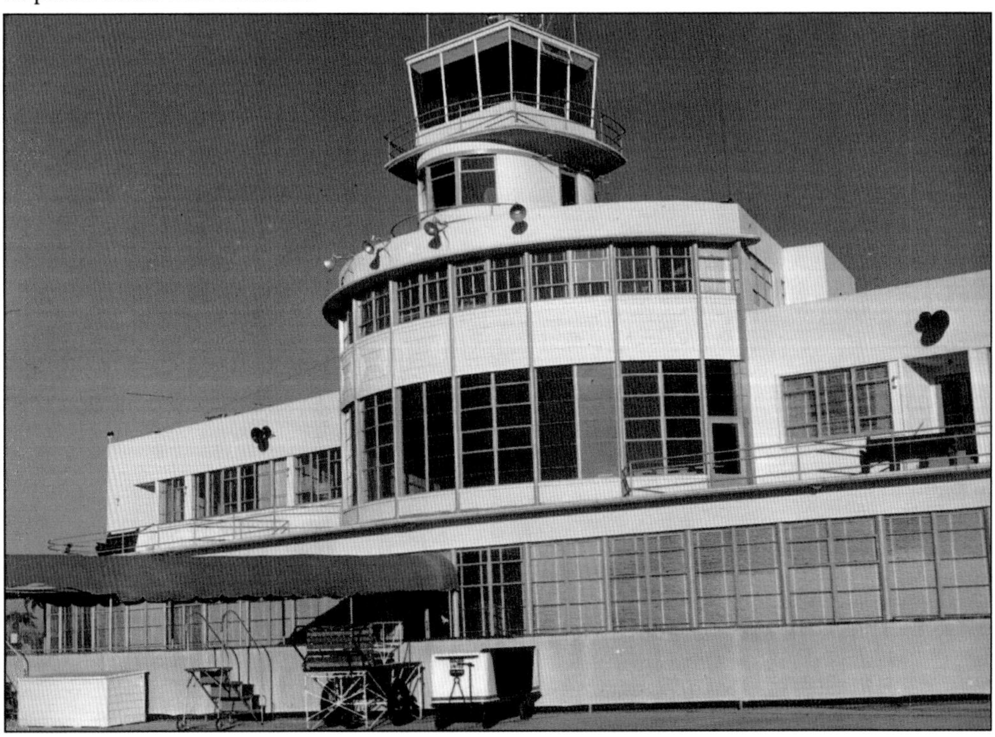

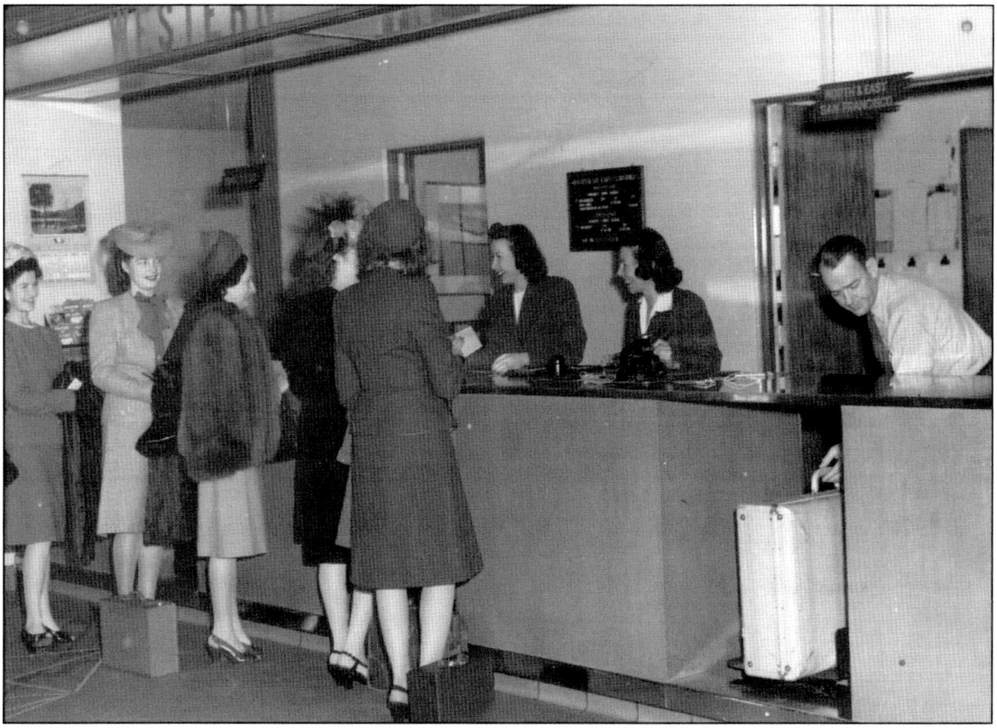

TICKETING AND RESERVATIONS. The first floor extended the length of the building and included airline offices where telephoned reservations were taken, a coffee shop, telephones, and waiting rooms facing the field. During the war, non-essential commercial air routes were abandoned. Western Airlines transported ferrying pilots and other military personnel to the field. By 1941, there were 174 non-military airports in California. Even during war time, Long Beach's future as a major airport for commercial airlines was threatened by the efforts of Los Angeles to wean air transport lines away from using the Long Beach airport when the war ended.

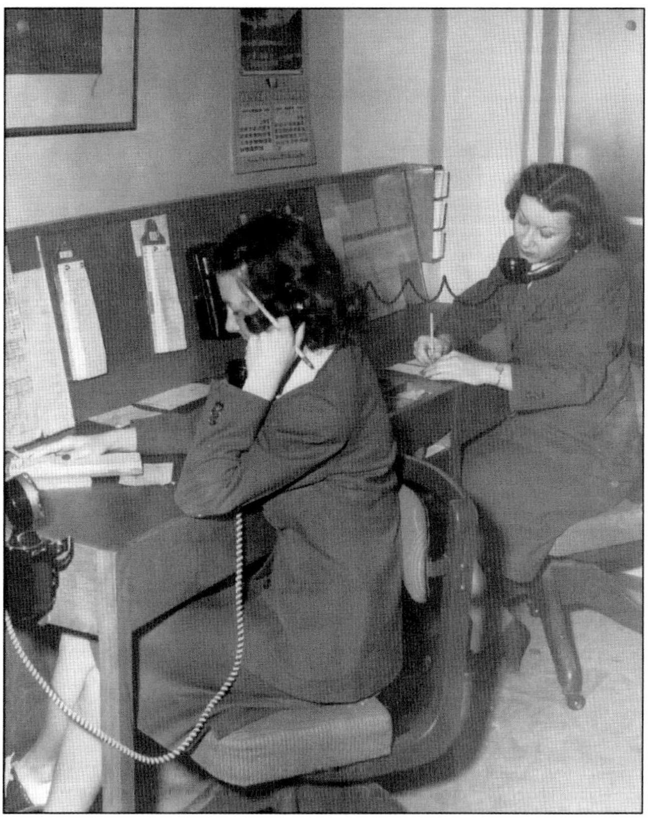

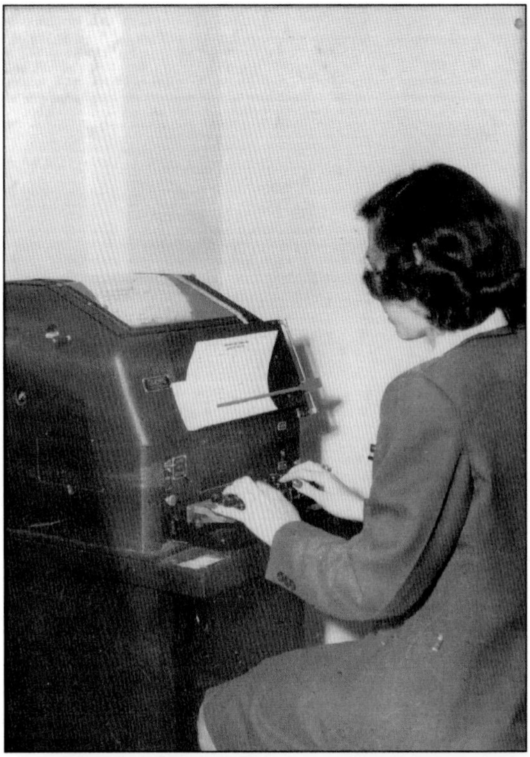

CONTROL TOWER. The second floor featured a large dining room (closed during the war), kitchen, lounge, and open deck overlooking the airfield. The third floor contained general airport administration, navigation instruments, radio range station, U.S. Weather Bureau, and teletype (left). Air traffic control (below) operated by the Civil Aeronautics Administration was on the "fourth floor." The terminal building formally opened April 26, 1942. By 1943, some 360,000 civilians and military personnel had patronized the terminal building, largely because of the coffee shop. The airfield, usually free of cloud cover, served as a backup for major airlines—Western, United, American, and TWA—when they could not land elsewhere. The city's airport manager, Earl Lloyd, visited other airports to determine how Long Beach could become "the West's greatest airport" after the war. Lloyd pushed for Long Beach to expand its 6,900-foot runway to 10,000 or 14,000 feet as he had seen in New York.

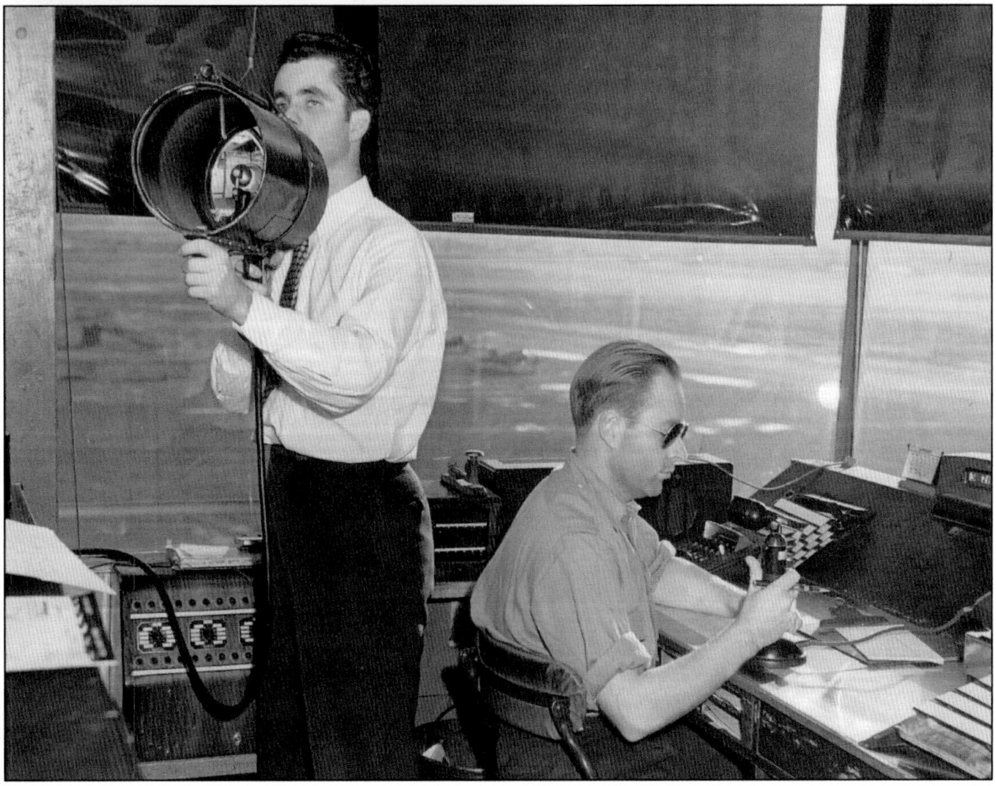

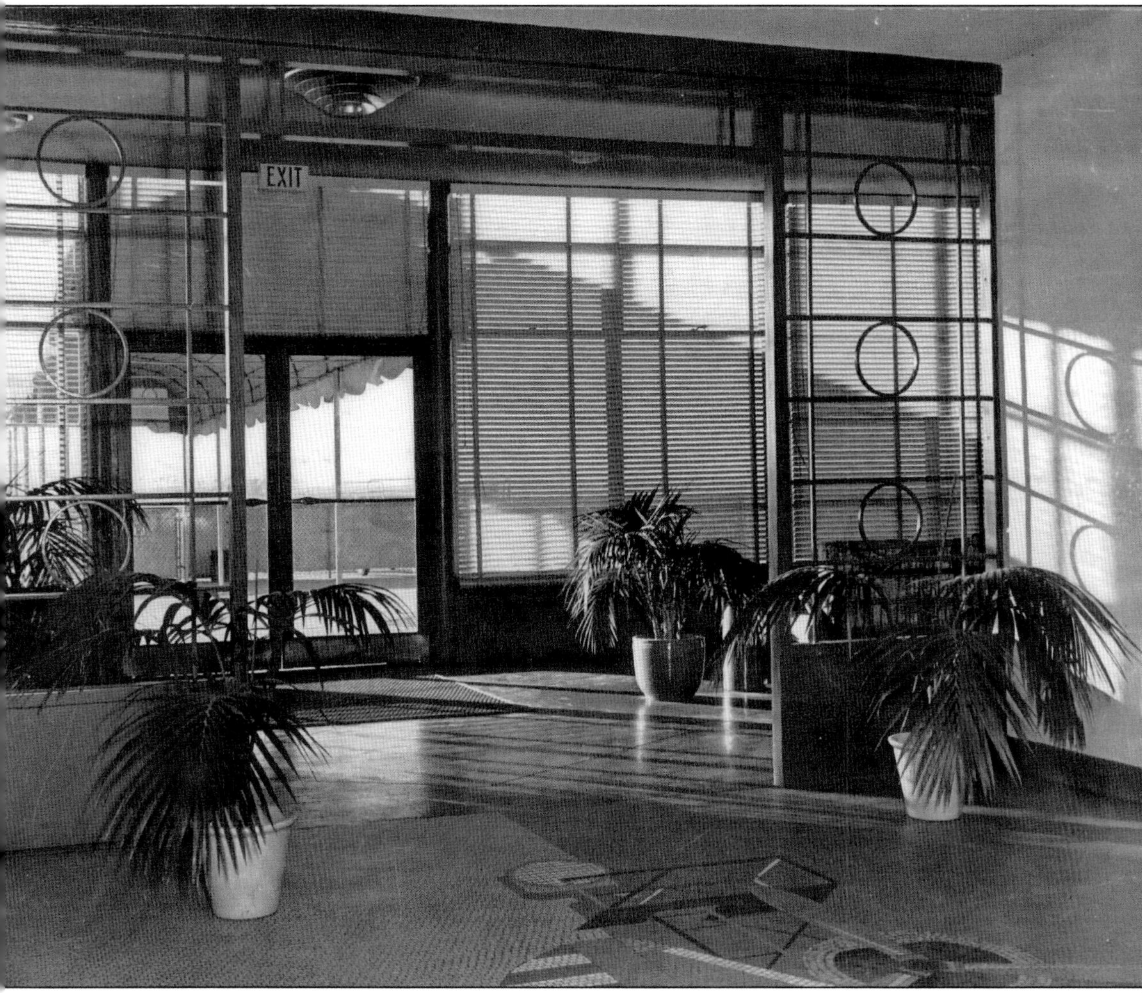

Grace Clements. Floor mosaics and wall murals inside the terminal building were designed as part of the Southern California Art Project of the Works Projects Administration (WPA) under the supervision of Grace Clements (1905–1969), "official supervisor of decorations of administration buildings at Daugherty Field." The December 1942 article "Murals and Mosaics by Grace Clements for the Long Beach Municipal Airport," in *California Arts and Architecture*, describes how Clements focused upon the theme of communication: "A large map of the western hemisphere showing air routes occupies the central portion of the concourse floor. Large geometric areas of unbroken color form the main body of the floor, highlighted by design units evolved from the idea of communication—ships, oil, aviation [shown in the picture] and telephone." Clements used assistants to lay the tile. The mosaics covered the 4,300-square-foot first floor. She also painted four murals on the first floor level dealing with "a particular means of communications—by land, by water, by air and by sound."

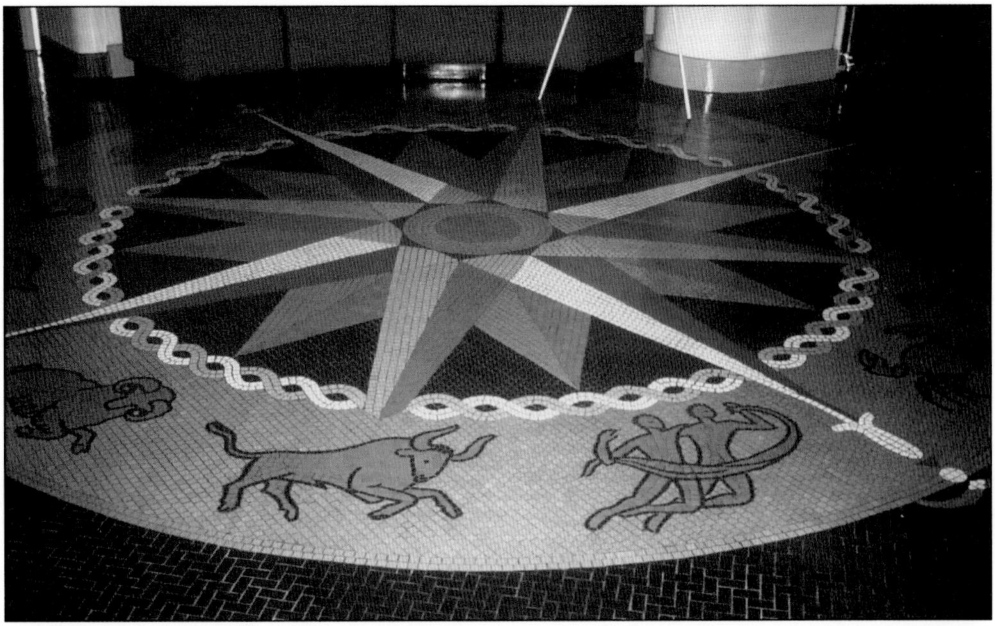

Zodiac Mosaic. Second-floor murals included sky and constellations of the northern hemisphere. A zodiac mosaic is outside the entrance of the restaurant (above). The colors in the murals were specifically designed to mute the strong light coming in from the windows in the dining room. All murals were painted over as of 2005. The picture below shows the concourse and one of the murals that was painted over. The mosaic city seal at the main entrance and seagulls at the back remain. Carpeting covers the three mosaics on the first floor, depicting oil, shipping, and aviation. Upstairs a tiled corridor and the zodiac mosaic can still be viewed.

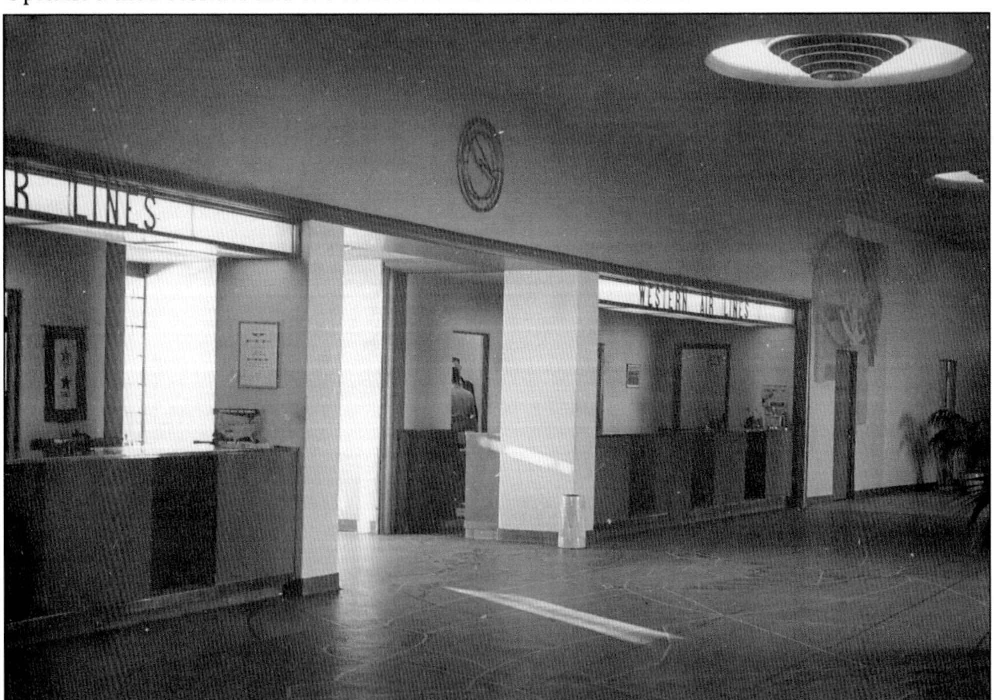

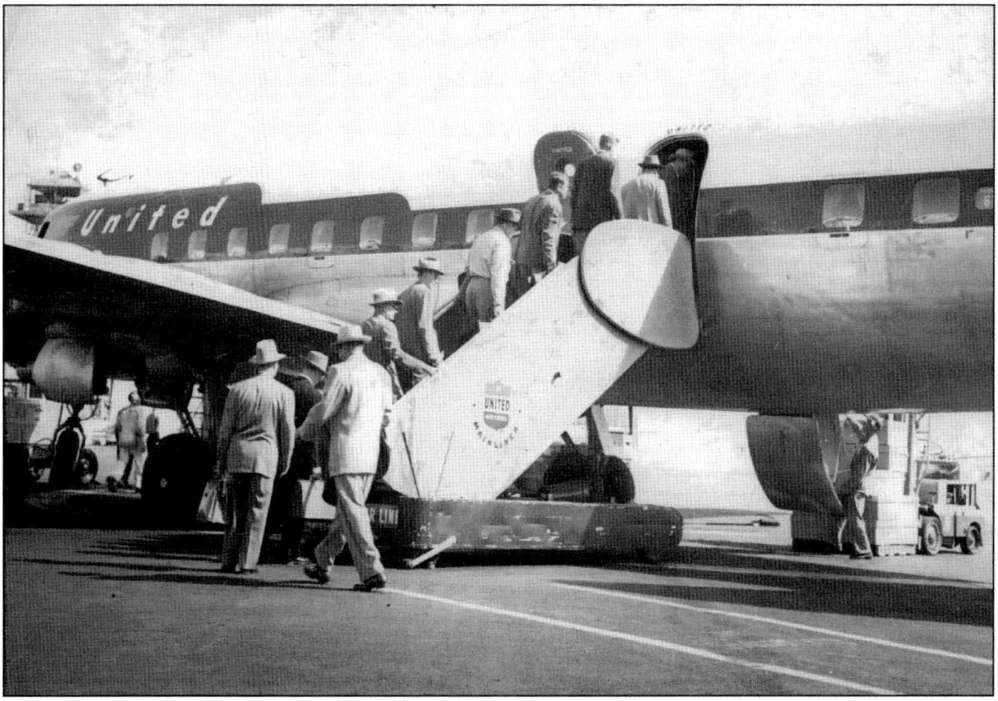

UNITED AIRLINES. By 1942, Long Beach Municipal Airport was "Number One in the Nation" because of its military and civilian aircraft activity. City officials recognized that once the war ended, the airport needed to expand to the east, which required state approval to close Lakewood Boulevard. On the south, the military structures needed removal. Los Angeles aggressively pushed "a super airport," opening a terminal in 1946. Long Beach voters rejected a bond ("Help Make Long Beach the Heart of Commerce") in 1945 to fund runway repairs and purchase land for expansion. Camouflage was removed from the terminal, and city officials announced, "The airport is ready to shift to full peacetime operations just as soon as the army moves away." It took more than 10 years for the military to leave. On July 3, 1946, United Air Lines (which had a ticket office in the terminal since 1940) made its first scheduled flight from the Long Beach Municipal Airport. Today passengers still walk out to the tarmac to board their flights, as they did in the 1940s.

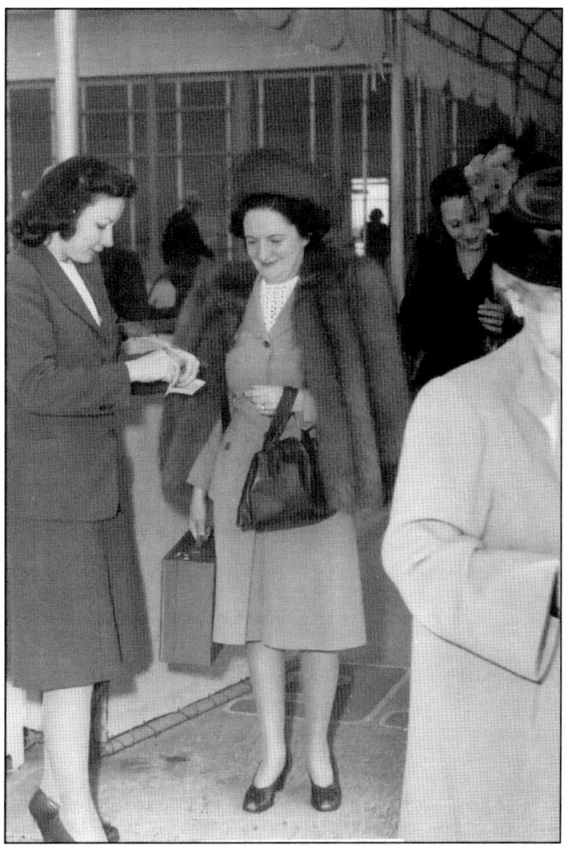

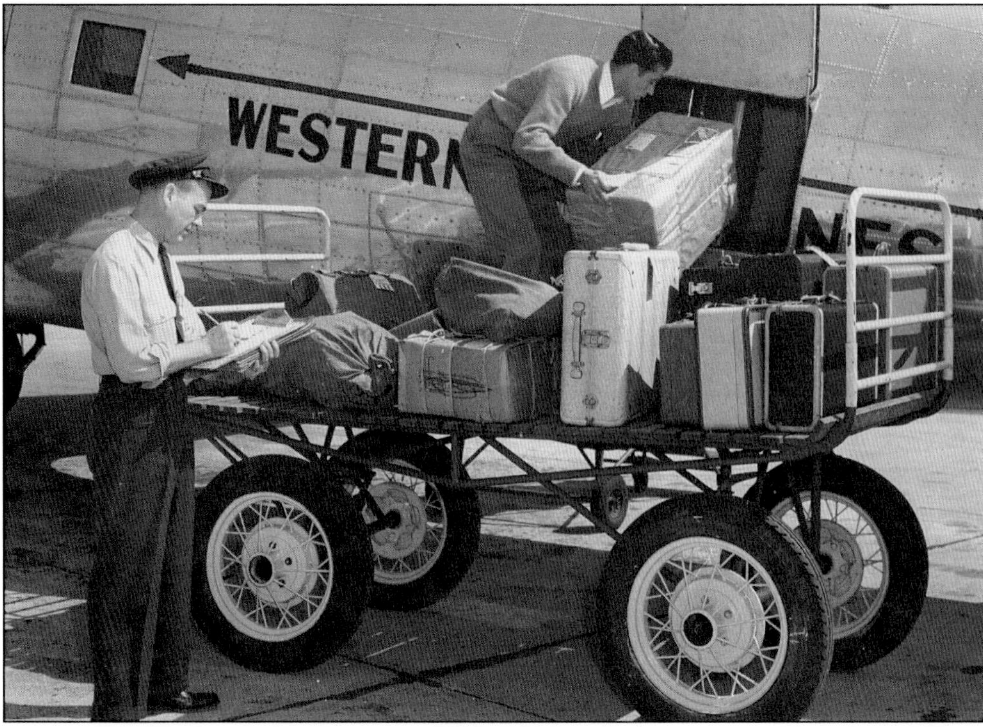

Passenger Travel vs. Air Cargo. In 1947, Long Beach airport chief Earl Lloyd resigned following criticism of his statements that Long Beach needed to decide whether it wanted to develop primarily for passenger travel or air cargo. He also called for the immediate cancellation of military leases. Local press noted, "Long Beach's prospects as an air traffic terminal are being injured every day's delay in putting this city in a position to compete actively with other municipalities for success in air transportation." The city explored the possibility of an airport off the waterfront near the breakwater because of the high cost of land near Daugherty Field. A radar landing system was installed on the field the same year, and residents of the Cal Heights–Los Cerritos area intensified complaints about their homes and schools being endangered by the army planes flying overhead.

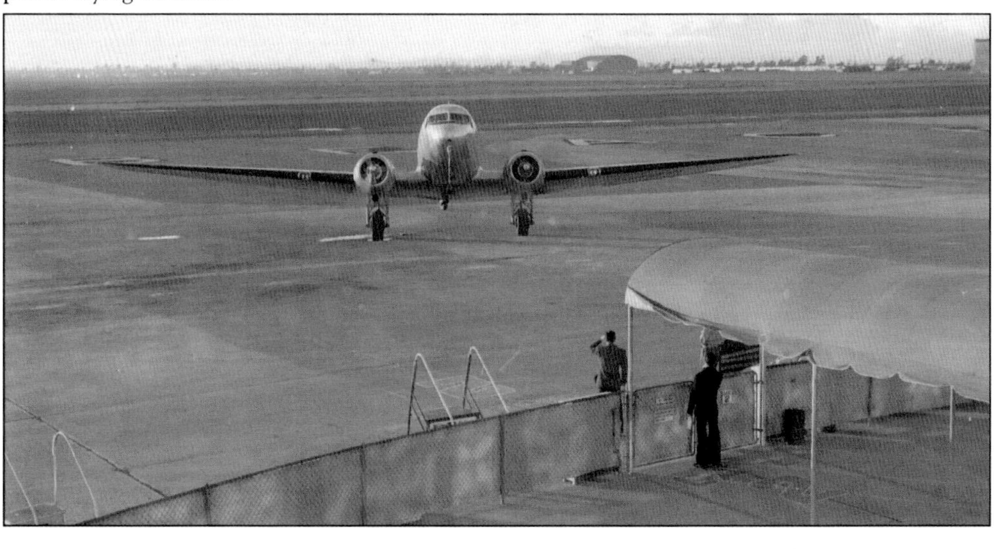

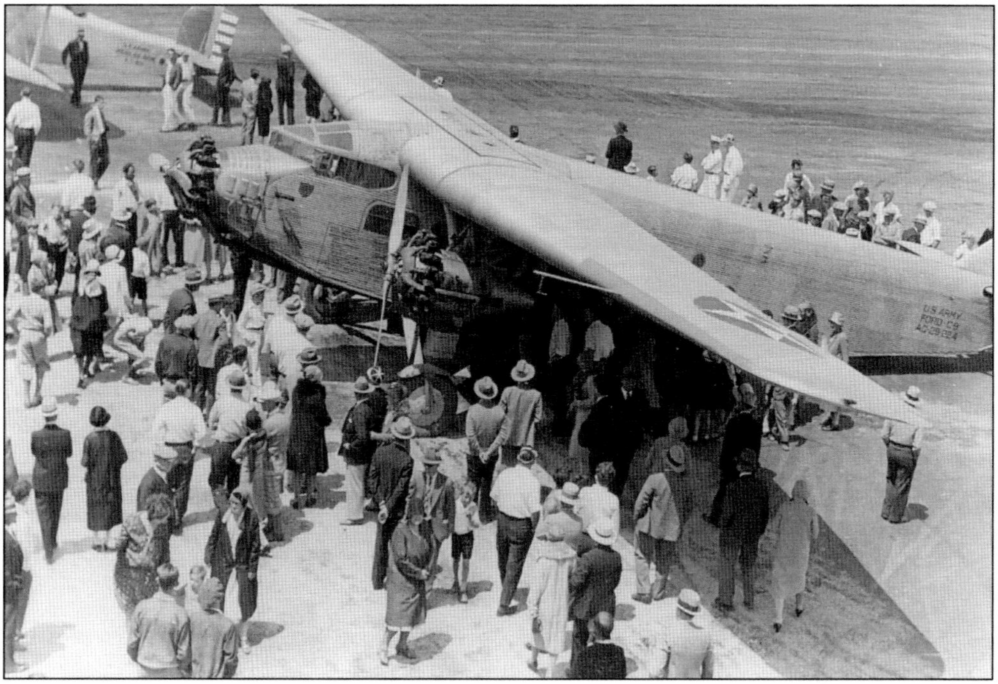

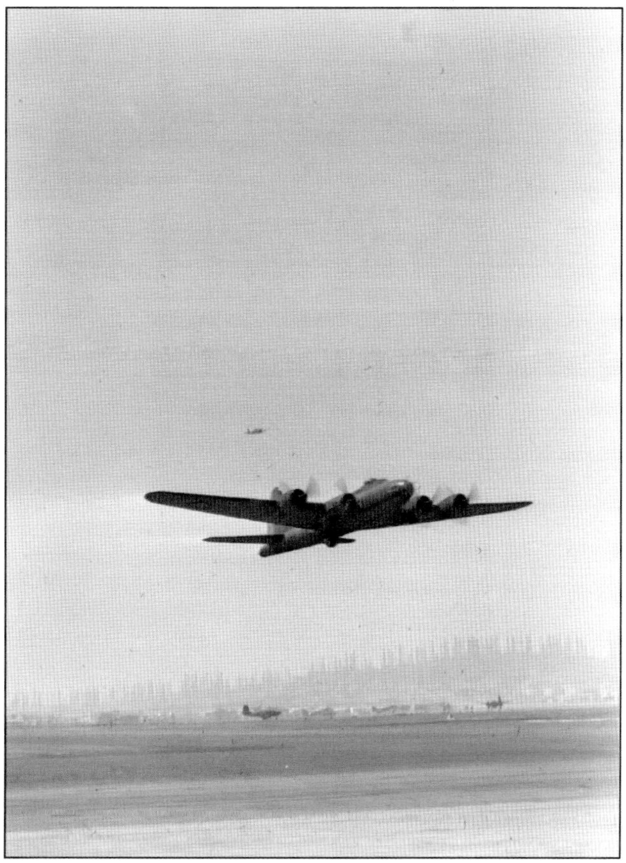

Post-Military Airport. The airport was a major attraction in Long Beach. While Long Beach was proud of its support and contribution to military aviation (such as the B-17 *Memphis Belle* pictured taking off from Daugherty Field), the city aggressively pushed the military out to make room for commercial aviation and to pursue drilling the oil and gas deposits underneath the field. By 1948, the airport handled 27,572 landings and takeoffs monthly. Of these, 300 were scheduled passenger airline flights. Two major factors kept Long Beach from being the terminus for the large airlines: the high amount of non-scheduled private planes flying in and out of the airport and the rapid development of residential housing adjacent to the airport, prompting the Civil Aeronautics Association to refer to Long Beach in 1948 as the "second stage of development to Los Angeles." (Both, BC.)

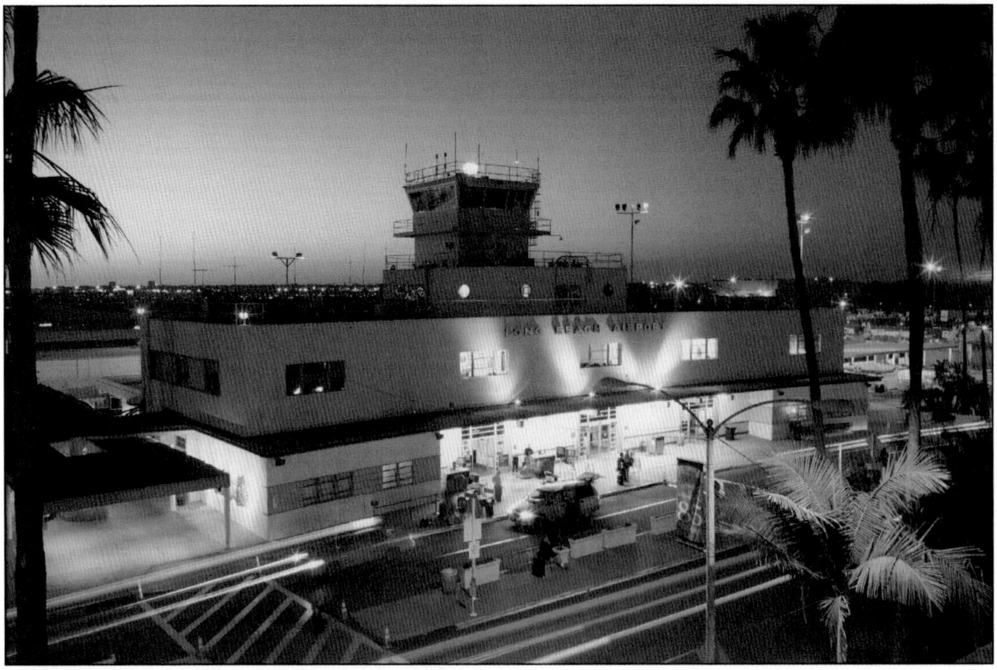

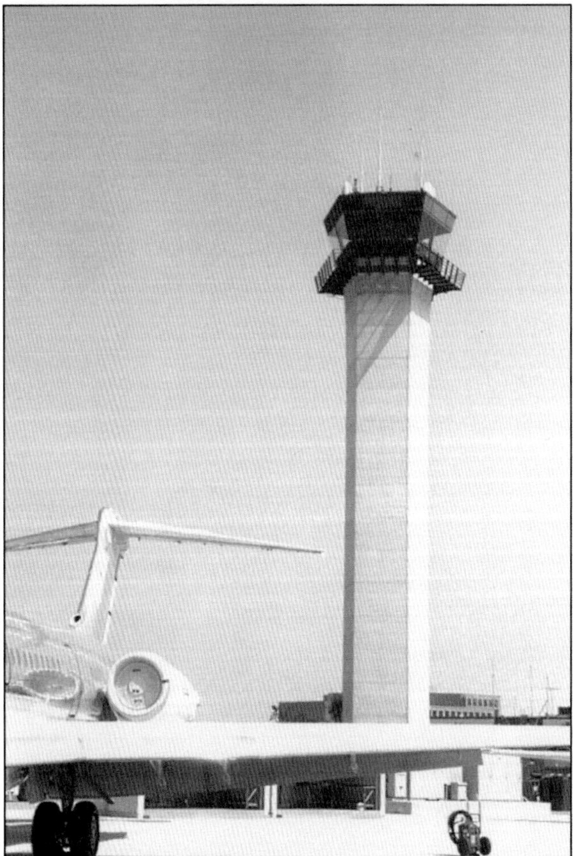

LONG BEACH MUNICIPAL AIRPORT–DAUGHERTY FIELD TODAY. Most of the California airports begun in the golden age of aviation have closed, but Long Beach Municipal Airport continues as one of the world's busiest general aviation airports. Its five runways, including one 10,000 feet, also accommodate airline and corporate passengers and air cargo services. The terminal building looks essentially as it did when built in 1941. The control tower was moved in 1971 by the Federal Aviation Administration from the top of the terminal to a free-standing structure south of the airfield on Spring Street. A new concourse and pre-boarding lounge south of the terminal were added in 1984. The city designated the terminal a Cultural Heritage Landmark in 1990. A noise ordinance enacted in 1981 limits the number of daily flights to 41 commercial and 25 commuter. In 2007, the city council approved a plan to improve the terminal and facilities. Long Beach was the first airport to install "solar trees" that provide 15,000 kilowatt hours to airport operations.

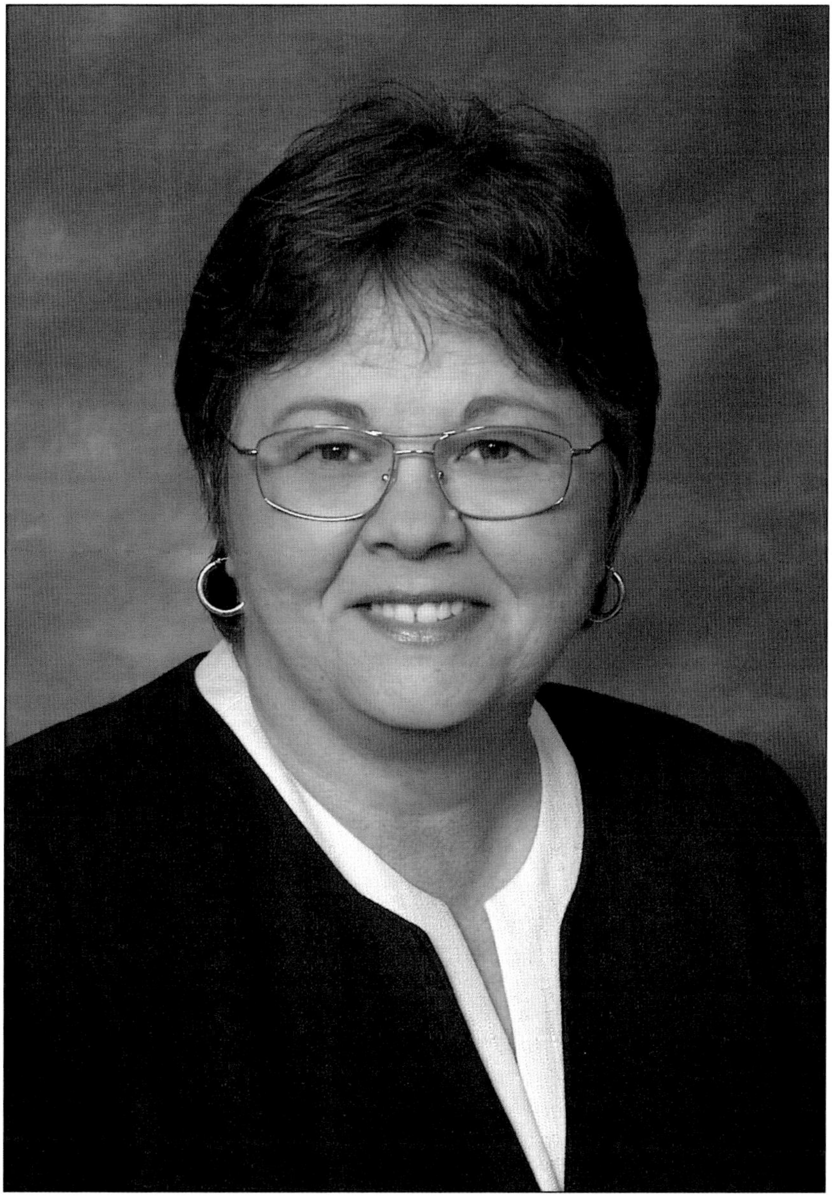

The Author. Following World War II, Gerrie Schipske's parents met at the Pike. Her father, Norman, a native of New Jersey, was an 18-year-old U.S. Marine assigned to the USS *Topeka*, docked in Long Beach. Her mother, Mary, a 17-year-old transplant from Pennsylvania, came with her family so that her father could build houses. Married at St. Lucy's Catholic Church on the west side of the city, her parents settled at Truman Boyd Manor, a housing complex named in honor of a World War II soldier. Gerrie was born in 1950 at the Long Beach Naval Hospital. She is a registered nurse practitioner and attorney at law. She taught women's studies and political science at California State University, Long Beach. In 1992, she was elected to the Long Beach Community College Board of Trustees. In 2006, she was elected to the Long Beach City Council and represents the 5th District, which includes the Long Beach Municipal Airport–Daugherty Field. She serves as president of the Long Beach Rosie the Riveter Foundation. She is also the author of *Rosie the Riveter in Long Beach*, an Arcadia publication. (Official photograph.)

Discover Thousands of Local History Books Featuring Millions of Vintage Images

Arcadia Publishing, the leading local history publisher in the United States, is committed to making history accessible and meaningful through publishing books that celebrate and preserve the heritage of America's people and places.

Find more books like this at
www.arcadiapublishing.com

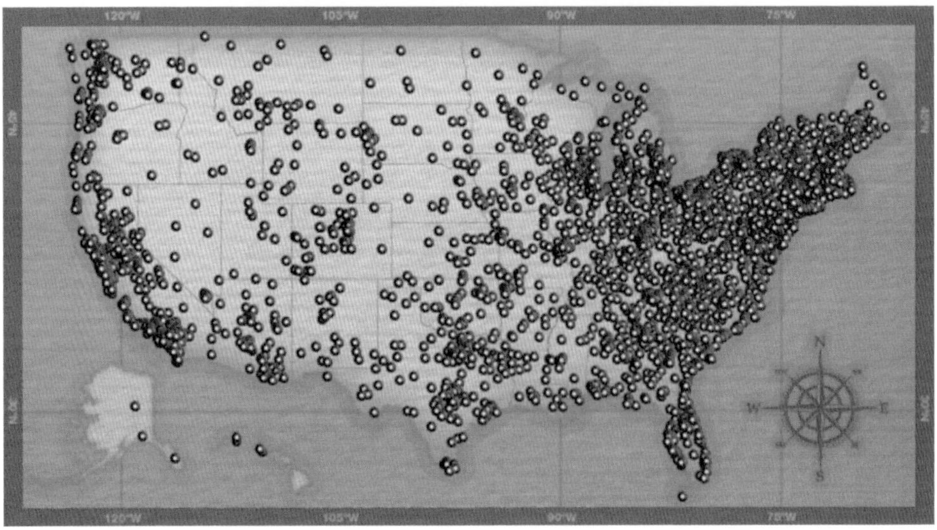

Search for your hometown history, your old stomping grounds, and even your favorite sports team.

Consistent with our mission to preserve history on a local level, this book was printed in South Carolina on American-made paper and manufactured entirely in the United States. Products carrying the accredited Forest Stewardship Council (FSC) label are printed on 100 percent FSC-certified paper.

MADE IN THE USA